Cosmology

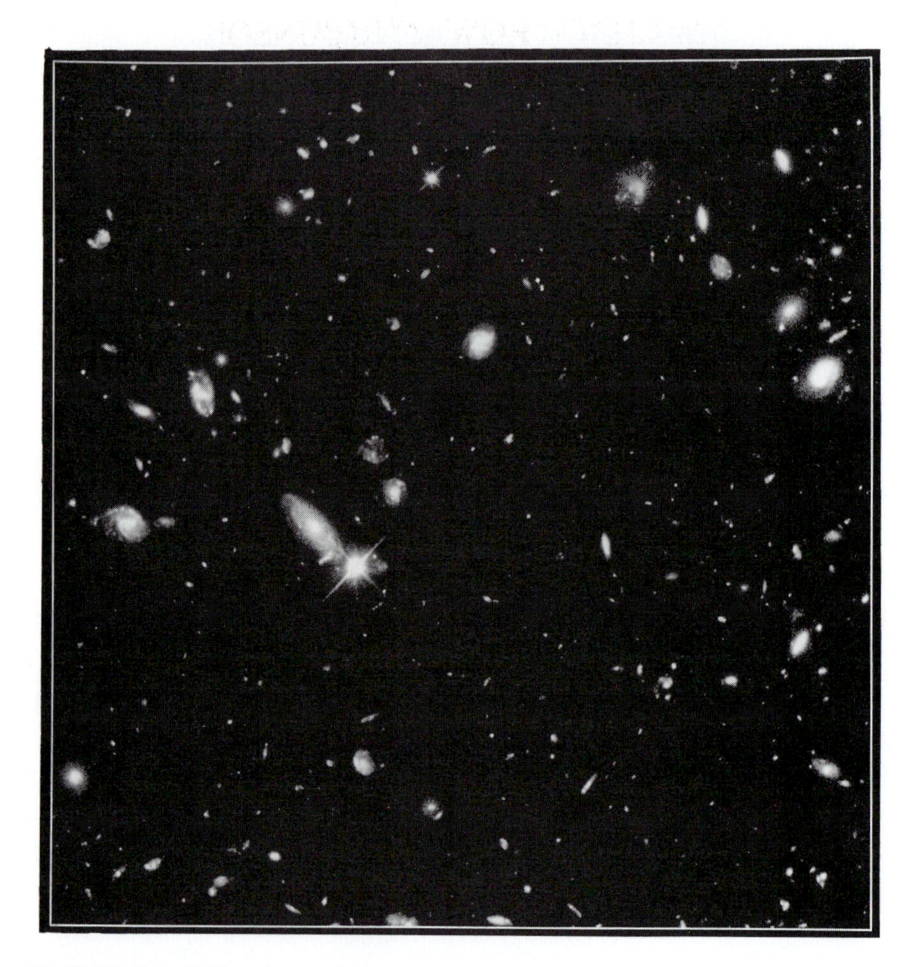

The Hubble Deep Field, an image of the extragalactic sky made from 276 frames from the Wide Field Planetary Camera of the Hubble Space Telescope. This is the deepest ever image of galaxies, many of which have redshifts > 1.

Cosmology

MICHAEL ROWAN-ROBINSON

Blackett Laboratory
Imperial College, London

THIRD EDITION

Clarendon Press · Oxford

Oxford University Press, Great Clarendon Street, Oxford OX2 6DP
Oxford New York
Athens Auckland Bangkok Bogota Bombay Buenos Aires Calcutta
Cape Town Chennai Dar es Salaam Delhi Florence Hong Kong Istanbul
Karachi Kuala Lumpur Madrid Melbourne Mexico City Mumbai
Nairobi Paris São Paolo Singapore Taipei Tokyo Toronto Warsaw
and associated companies in
Berlin Ibadan

Oxford is a trade mark of Oxford University Press

Published in the United States
by Oxford University Press Inc., New York

First published 1977
Second edition 1981
Third edition 1996
Reprinted 1998

A catalogue record for this book is available from the British Library

Library of Congress Cataloging in Publication Data

Rowan-Robinson, Michael.
Cosmology / Michael Rowan-Robinson. – 3rd ed.
Includes bibliographical references and index.
1. Cosmology. I. Title
QB981.R69 1996 523.1–dc20 96–1769

ISBN 0 19 851885 4 (Hbk)
0 19 851884 6 (Pbk)

Printed in Great Britain
by Bookcraft Ltd,
Midsomer Norton, Nr Bath, Avon

Preface to the third edition

When I started to write the first edition of this book in 1972, the microwave background radiation had been discovered only a few years previously. Although the big-bang models had begun to seem quite well established, cosmology remained a highly controversial subject. Today there is little room for doubt about the correctness of the big-bang models.

In the almost quarter of a century since then, there have been enormous strides both in our observations of the large scale and in our theoretical understanding of the universe. It has been a major task to revise the book to take account of the discoveries of the 16 years since the second edition and I have had to start again on several occasions. The impact of space missions like IRAS, COBE, and the Hubble Space Telescope, and of the major new ground-based telescopes, has been immense. We seem to be close to understanding the nature of dark matter in the universe and how galaxies formed, and to measuring key cosmological quantities like the Hubble and density parameters. We are also able to speculate meaningfully about the earliest instants of the big bang.

Yet cosmology remains in an exciting and controversial state, and a key section of this book is the Epilogue: Twenty controversies in cosmology today. Only a few of the headings in this section, and almost none of the text remain the same as in the first edition. The continuing capacity of the universe to surprise us and to maintain its enigma is what makes cosmology such a fascinating subject.

I think I was the first in the UK, in 1968, to give a course on cosmology at an undergraduate level. Now almost all departments of physics or applied mathematics give such a course (and if they don't they should!). As well as bringing the text up to date, I have expanded the problems and provided solutions. Several eminent young cosmologists have told me they first started to learn cosmology from earlier editions of this book. I hope this edition will introduce the wonderful subject of cosmology to another generation.

London
March 1996

M. R.-R.

Preface to the first edition

From the earliest times people have asked questions about the universe they find themselves in. We are fortunate to be alive in the third great age of cosmology. The first was the age of Epicurus, Aristarchus, and Hipparcos, the third and second centuries BC, when the notion of an infinite universe in which the Earth is not at the centre was first considered, though not adopted. The second began with the methodical programme of Copernicus to prove the motion of the Earth and ended with that great cry of Bruno: 'The stars are suns like our own and there are countless suns freely suspended in limitless space, all of them surrounded by planets like our own Earth, peopled with living beings. The Sun is only one star among many, singled out because it is so close to us. The Sun has no central position in the boundless infinite.'

In the third age of cosmology, which could be said to have begun with Einstein's proposal of an isotropic and homogeneous universe, we are seeing the whole electromagnetic spectrum, from radio through microwave and infrared to X- and γ-rays, pressed into the service of cosmology. We are probably only at the beginning of the revolution that these new windows on the universe will bring about.

Despite the wealth of new information, the models of the universe that are now favoured are the simplest big-bang models put forward by Einstein, de Sitter, Friedmann, and Lemaître in the 1920s. The decisive factor in this convergence of thinking has been the cosmic microwave radiation discovered by Penzias and Wilson in 1965. For no explanation of this other than that it is the relic of the fireball phase of an isotropic big bang has stood the test of time. The steady-state cosmology has long since fallen by the wayside.

This therefore seems a good moment to write a book outlining modern ideas about cosmology at an elementary level. The book appears in a series aimed at first-year maths and physics undergraduates, but I hope that most of it will be accessible to those who have studied science to an advanced level at school.

Because this is neither a research text nor a history of cosmology I have abandoned the practice, alas seldom strictly adhered to, of giving references to all sources of ideas and facts. The reader will easily see the extent to which modern science is a collective enterprise by consulting the review articles referenced at the end of this book, which between them refer to the literally thousands of scientific papers on which the picture presented here is based. I crave the indulgence of the many hundreds of colleagues who should have been acknowledged in this way. They at least will know the part they have played in this book.

As I remain deeply sceptical of many of the ideas outlined here, I have tried to emphasize the open-endedness of cosmology, the fact that the whole argument is still going on. To underline this I have included an epilogue outlining twenty controversies in cosmology today.

I must record my thanks to Professor W. H. McCrea, both for his lectures on cosmology which first aroused my interest in the subject and for reading and commenting on the manuscript of this book. My thanks also to several colleagues and friends who read the manuscript, made suggestions, and corrected errors, especially Wal Sargent, Laura Maraschi, Ian Roxburgh, Paul Davies, and Andy Fabian, but they are in no way responsible for the errors and deficiencies that remain. I owe a special debt of gratitude to Professor R. L. F. Boyd for his encouragement and interest in the book from its earliest stages and for his careful reading and criticism of the first draft.

London M. R-R.
January 1975

Preface to the second edition

In the five years since the first edition was completed there has been intense activity in almost all areas of cosmology and several hundred new papers have been written on the topics covered in the book. I have therefore revised many parts of the book and expanded it slightly.

The sections on helium production and on anisotropic and inhomogeneous models have been expanded, and there are new sections on the isotropy and spectrum of the microwave background, on the evolution of density fluctuations and the formation of galaxies and clusters, on grand unified theories, and on cold and tepid big-bang models. The discussions of observations have been brought up to date, particularly those relating to X- and γ-ray astronomy, to clusters of galaxies, and to experimental gravitation.

The over-all big-bang picture is now more secure than ever, but there are still many areas of controversy. There were only a few of my 'Twenty controversies in cosmology today' which I felt had been completely resolved, although the area of debate had shifted on many of them.

I have of course taken the opportunity of this new edition to correct the errors that have been pointed out to me, and I particularly thank George Ellis, Joseph Silk, John Barrow, and Malcolm MacCullam for their detailed comments.

January 1980 M. R-R.

Acknowledgements

I thank the following for permission to use photographs and figures: NASA: Cover, frontispiece, Figs 1(a), 5.1(b); California Institute of Technology: Figs 2.1(b), 2.4(b),(d), 2.7(a), 2.9; W. B. Saunders: Fig. 1.4; Y. Suzuki: Fig. 1.8(a); J. Bechtold: Fig. 1.10(a); Mount Wilson Observatory: Fig 2.1(a); Lick Observatory: Fig. 1.3, 2.4(a),(c),(e); P. C. van der Kruit: Fig. 2.7(b); National Radio Astronomy Observatory: Fig. 2.8(a); G. Hasinger: Fig. 2.10; Royal Observatory Edinburgh: Fig 2.12; University of Chicago Press: Figs 3.4, 6.5(b); C. D. Shane and Lund Observatory: Fig. 3.5; G. P. Eftsathiou: Fig. 3.6(a); J. J. Condon: Fig. 3.6(b); T. Shanks: Figs 3.7, 7.11; A. Penzias: Fig. 5.1(a); D. Wilkinson: Fig. 5.1(c); M. Turner: Figs 6.1, 6.2; G. Borner: Fig. 6.3; T. S. van Albada: Fig. 6.5(b); A. Sandage: Fig. 7.4; J. S. Dunlop: Fig. 7.5; K. Kellerman and Macmillan Journals Ltd: Fig. 7.9.

I thank all those colleagues who have commented on earlier editions, especially Malcom MacCallum and Bernard Carr, and Stephen Warren for his careful reading of the manuscript for this edition.

Contents

This world has persisted many a long year, having once been set going in the appropriate motions. From these everything else follows.

LUCRETIUS

Do you want to stride into the infinite?
Then explore the finite in all directions.

GOETHE

And I say to any man or woman. Let your souls stand cool and composed before a million universes.

WHITMAN

1
The visible universe

1.1 Introduction

Imagine that it is a clear moonless night in the country. The sky is ablaze with thousands of stars. The light from some of the faintest ones set out thousands years ago.

In a great arc across the sky we see that familiar band of light, the Milky Way, which Galileo showed to be composed of myriads of faint stars. This is our own Galaxy, a discus-shaped metropolis of a hundred thousand million (10^{11}) stars with ourselves far out towards the suburbs (see Fig. 1.1).

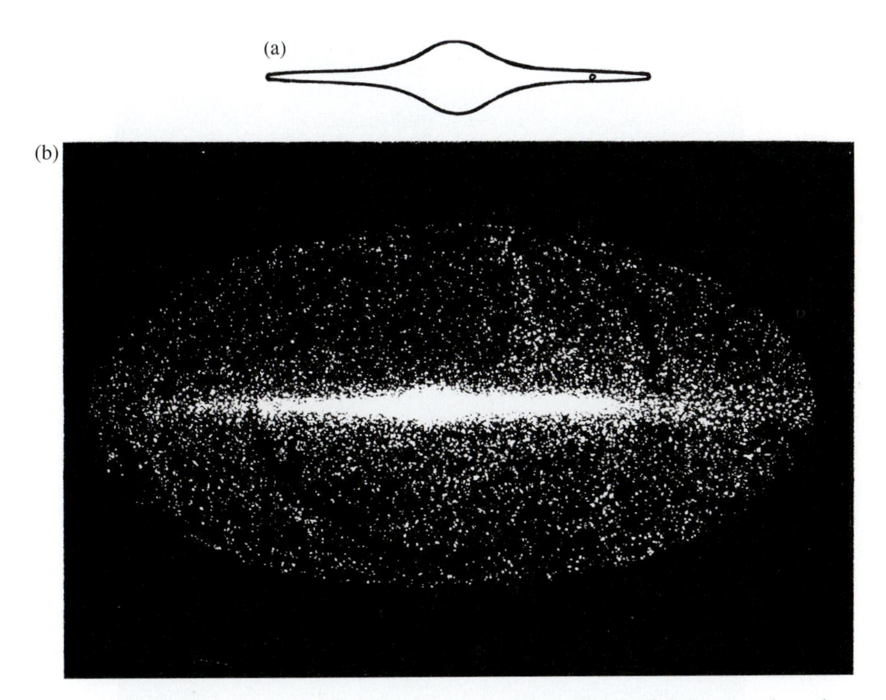

Fig. 1.1 (a) The outline of our Galaxy as it would look edge-on. The small circle, centred on where the sun would be, indicates the region where most of the stars visible to the naked eye lie. The Milky Way consists of the integrated light from more distant stars in the disc. (b) Stellar sources, mainly red giants, seen in the infrared by the Infrared Astronomical Satellite (IRAS), delineating the disc and the bulge of our Galaxy.

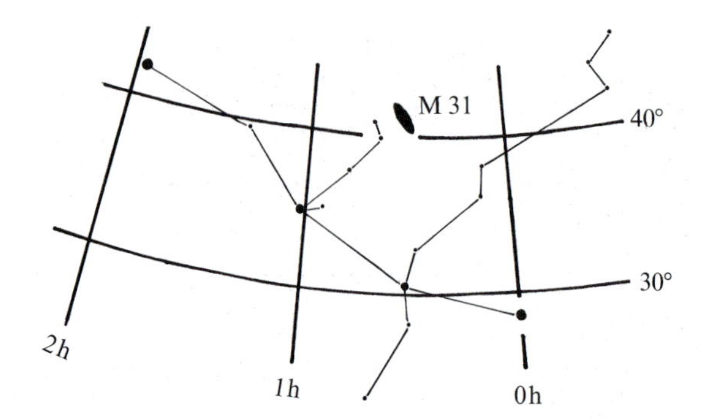

Fig. 1.2 The constellation of Andromeda showing the location of the nebula Messier 31. In 1924 Hubble showed that M31 lies far outside our Galaxy. The coordinates shown are right ascension (horizontal axis) and declination (vertical axis), which are equivalent to longitude and latitude projected on to the sky.

Fig. 1.3 The Andromeda nebula, a spiral galaxy seen tilted to our line of sight. Messier 31 and our own Galaxy are the dominant members of the Local Group of galaxies. Two other members, the dwarf elliptical galaxies NGC205 and 221, can also be seen in the photograph, which is from Lick Observatory.

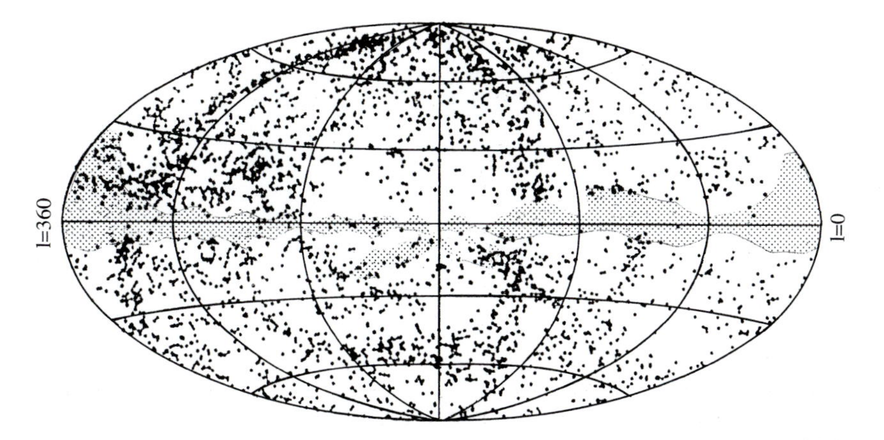

Fig. 1.4 The distribution on the sky of galaxies with optical diameters greater than 2 arcmin. The concentration to the upper left is the Virgo cluster. Shaded area is obscured by galactic dust.

As recently as 1920 it was still reasonable to believe that this gigantic star system, 60 000 light years in diameter (1 light year = distance travelled by light in one year = 0.946×10^{16} m) comprised the whole visible universe. Today our horizon is at least 300 000 times larger.

As a first step out from our own Galaxy let us, with the aid of a large telescope, travel towards a faint and fuzzy patch of light in the constellation of Andromeda, the nebula M31 (see Figs 1.2 and 1.3). Two million lights years away, it is almost the twin of our own Galaxy, seen tilted to our line of sight. These two galaxies are two of the dominant members of a small group of 30 or so galaxies known as the Local Group of galaxies (see Table 1.1).

Now let us travel 60 million light years towards the constellation of Virgo. We find ourselves in a cloud of thousands of galaxies, the Virgo cluster, first recognized by William Herschel (see Fig. 1.4). Our Galaxy may lie in the outer fringes of this cluster.

Suppose we travel out to the limit of vision of the 10 m Keck telescope on Mauna Kea, Hawaii. The most distant galaxies we can see are at least 10^{10} light years away. Their light set out long before the Earth was formed. This whole expanse is filled with galaxies and clusters of galaxies. This 'realm of the nebulae', as Hubble called it (see references, p. 157) is the subject of this book.

1.2 The electromagnetic spectrum

Since even the nearest star, α Cen, 4 light years away, lies for the moment far beyond our reach, we can learn about distant parts of the universe only from the light and other kinds of information they send us.

The human eye responds to only a very narrow range of frequencies, the *visible* portion of the electromagnetic spectrum (Fig. 1.5). In 1800 Herschel first showed the

Table 1.1 The Local Group of galaxies

Name	Type (see p. 31)	Distance (10^3 light years)	lg (mass of galaxy/mass of sun, $M/M_{\odot}$)	Absolute visual magnitude (see p. 47)	Linear diameter (10^3 light years)	Radial velocity (km s^{-1})
M31	Sb	2200	11.5	–21.1	50	–275
Our galaxy	Sab?	26 (to centre)	11.2	–20.5	80	0
M33 (NGC598)	Sc	2400	10.1	–18.8	20	–190
Large Magellanic cloud	Irr	170	10.0	–18.7	24	–270
NGC205	E5	2100	9.9	–16.3	6	–240
M32 (NGC221)	E2	2200	9.5	–16.3	3	–210
Small Magellanic cloud	Irr	210	9.3	–16.7	10	168
NGC147	E_{pec}	2200	9	–14.8	3	–250
NGC185	E_{pec}	2200	9	–15.2	3	–300
Pegasus	Irr	7500		–16.7	8	–181
NGC6822	Irr	1500	8.5	–15.6	6	–40
IC10	Irr	4200		–15.3	6	–343
Wolf–Lundmark	Irr	5200		–15.0	9	–78
IC1613	Irr	2400	8.4	–14.8	3	–240
IC5152	Irr	5200		$-14.3^{\dagger}$	6	78
Leo A	Irr	5200		–13.4	6	26
Fornax	E	550	7.3	–13	6	40
Leo I	E4	750	6.6	–11	3	
Sculptor	E	280	6.5	–12	3	
Leo II	E1	750	6.0	–9.5	3	
Draco	E	220	5	–8.5	1	
Ursa minor	E	220	5	–9	3	

Name	Type (see p. 31)	Distance (10^3 light years)	lg (mass of galaxy/mass of sun, $M/M_\odot$)	Absolute visual magnitude (see p. 47)	Linear diameter (10^3 light years)	Radial velocity (km s^{-1})
UKS1927 – 177	Irr	3600		–10.5†	3	
UKS2323 – 326	Irr	4200		–10.4†	2	
Sag DIG	Irr	1500	7.1	–9.3†	3	–58
LGS3	Irr	2400		–9.0†	2	–280
And I		2200		–11	2	
And II		2200			2	
And III		2200			3	
And IV		2200			1	

†Blue magnitude.

Other possible members include: DDO187, GR8, Sex A, Sex B, and NGC3109.

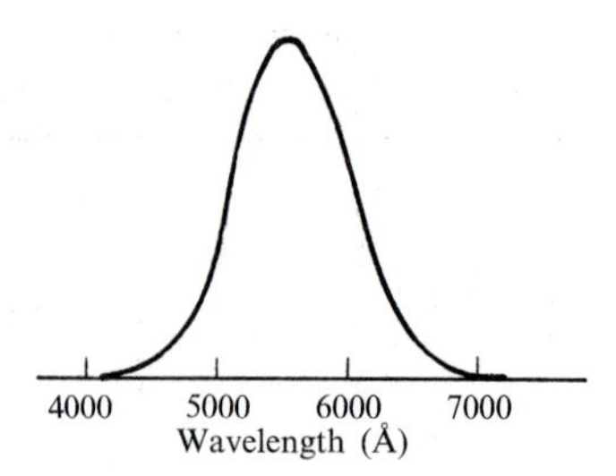

Fig. 1.5 The response function of the human eye, as a function of wavelength.

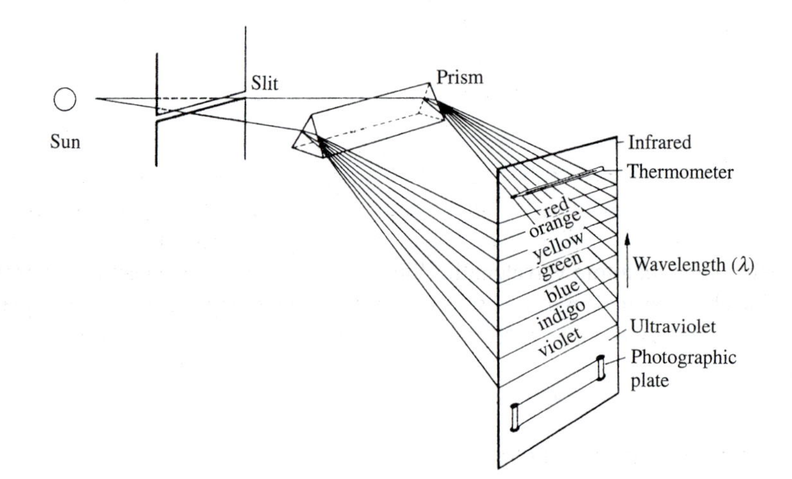

Fig. 1.6 Schematic illustration of the formation of a spectrum of the Sun's light by means of a prism.

possibility of using other frequencies or wavelengths for astronomy. He held a thermometer beyond the red portion of the spectrum obtained by passing the Sun's light through a prism, demonstrating the existence of *infrared* radiation from the Sun (Fig. 1.6). Soon afterwards Ritter found *ultraviolet* radiation. But it was not until 1931 that the American radio scientist Karl Jansky first showed that the Milky Way emitted *radio* waves. And in 1948 a rocket was launched carrying a photographic emulsion that recorded *X-rays* from the Sun. The detectors used to record these different types of radiation vary widely, from radio receivers and geiger counters to bolometers and photographic plates. And because many types of radiation are absorbed by the atmosphere, a great variety of observing platforms are used, from high-altitude observatories to aeroplanes, balloons, and satellites.

But it is very important to be clear that these different radiations are all fundamentally the same, namely *electromagnetic radiation*, or more simply 'light'. The wavelength λ and frequency ν are related by the equation

$$\lambda\nu = c \tag{1.1}$$

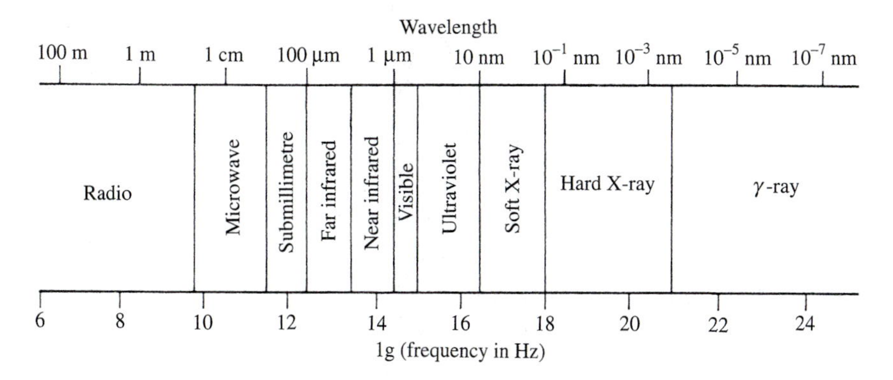

Fig. 1.7 The electromagnetic spectrum.

where c is the velocity of light. Figure 1.7 illustrates how the electromagnetic spectrum is arbitrarily divided into bands, defined primarily by the different detection techniques used. Radio, visible, and X-ray photons differ only in their frequency (and therefore wavelength). Since a photon carries an energy $h\nu$ where h is Planck's constant, X-ray photons carry far more energy than, for example, radio photons. Table 1.2 (p. 8) summarizes the electromagnetic spectrum for astronomy.

1.3 Astronomy without light

Electromagnetic radiation is not the only way that astronomical information reaches the Earth.

Cosmic rays

These energetic particles, electrons and the nuclei of atoms, moving at velocities very close to the speed of light, bombard the solar system continuously from all directions. While you are reading this sentence one will probably pass through your head. The kinetic energies of some of these particles far exceed anything achieved in the largest human-made accelerator (Fig. 1.8). On entry to the atmosphere, the higher-energy nuclei soon collide with molecules of air, creating a shower of secondary particles, which can be detected on the ground.

The Sun is an important source of low-energy cosmic rays. The main source of the more energetic particles is unknown, but it is believed that pulsars (pulsating radio sources associated with neutron stars), supernovae (explosions in high-mass stars that have exhausted their nuclear fuel), active galaxies (in which violent and explosive events of some kind are taking place), and quasars (quasi-stellar radio sources) all contribute.

Table 1.2 The electromagnetic spectrum for astronomy

Band	Radio	Microwave	Submillimetre	Far infrared	Near infrared
Frequency range	10^{6}–10^{10} Hz	10^{10}–$10^{11.5}$ Hz	$10^{11.5}$–$10^{12.5}$ Hz	$10^{12.5}$–$10^{13.5}$ Hz	$10^{13.5}$–$10^{14.6}$ Hz
Wavelength range	3 cm–300 m	1 mm–3 cm	100 μm–1 mm	10–100 μm	0.8–10 μm
Observing platform	Ground (satellite for $\mu < 3 \times 10^{7}$ Hz)	Ground	Mountain (1 mm, 300 μm) balloon, rocket, aircraft	Ground, aircraft, balloon satellite	Ground
Detector	Radio receiver	Radio receiver	Bolometer	Bolometer	Photograph, CCD
Bright observed sources	Cas A (supernova remnant) Sun Cyg A (radio galaxy) Sag A (galactic centre) Cen A (radio galaxy) Vir A (radio galaxy) Radio galaxies Quasars	Sun Crab nebula (supernova remnant) Omega nebula Cas A Orion nebula Radio galaxies Quasars	Sun Planets Galactic centre Orion nebula Gas and dust clouds in the Galaxy 3C273 (quasar) 3C84 (Seyfert galaxy) M82 (starburst galaxy)	Sun Planets Gas and dust clouds in the Galaxy Normal and Starburst galaxies	Sun Planets Cool stars or stars in dust clouds Galaxies Quasars
Faintest detectable flux (centre of band)	10^{-30} W m^{-2} Hz^{-1}	10^{-28} W m^{-2} Hz^{-1}	10^{-27} W m^{-2} Hz^{-1}	10^{-28} W m^{-2} Hz^{-1}	10^{-34} W m^{-2} Hz^{-1}
Approximate number of sources in sky to this level	10^{8}	10^{4}	10^{3}	10^{5}	10^{10}
Main contributing types of sources	Radio galaxies, quasars	Radio galaxies, quasars	Galactic sources	Stars, galaxies	Stars, galaxies
Main contribution to background radiation	Milky Way, radio galaxies	Cosmic, 2.7 K black body	Atmosphere	Atmosphere	Atmosphere, Zodiacal light

Band	Visible	Ultraviolet	Soft X-ray	Hard X-ray	γ-ray
Frequency range	$10^{14.6}$–$10^{14.9}$ Hz	$10^{14.9}$–$10^{16.5}$ Hz	$10^{16.5}$–$10^{17.5}$ Hz	$10^{17.5}$–10^{20} Hz	$> 10^{20}$ Hz
Wavelength range	0.4 cm–0.8 μm	100–4000 Å	10–100 Å (0.12–1.2 keV)	(1.2–370 keV)	(> 370 keV)
Observing platform	Ground	Ground (3000–4000 Å) Rocket, satellite ($\lambda < 3000$ Å)	Rocket, satellite	Balloon, satellite	Balloon, satellite
Detector	Photograph, image tube, CCD	Photograph, image tube		Photon counters	
Bright observed sources	Sun Planets Stars Galaxies Quasars	Sun Hot stars Galaxies Quasars	Sun Galactic sources (binary systems with white dwarf, neutron star, or black hole) Rich clusters of galaxies M82 (active galaxy) 3C273 (quasar)		Crab pulsar Galactic disc Vela pulsar 3C273 (quasar)
Faintest detectable flux (centre of band)	10^{-34} W m^{-2} Hz^{-1}		10^{-34} W m^{-2} Hz^{-1}		10^{-35} W m^{-2} Hz^{-1}
Approximate number of sources in sky to this level	10^{10}		10^{6}		30
Main contributing types of sources	Stars, galaxies		Galactic sources, clusters of galaxies, quasars,		
Main contribution to background radiation	Atmosphere, zodiacal light, scattered starlight (galaxies)		Seyferts quasars, Galactic gas, intergalactic gas?		Extragalactic sources?

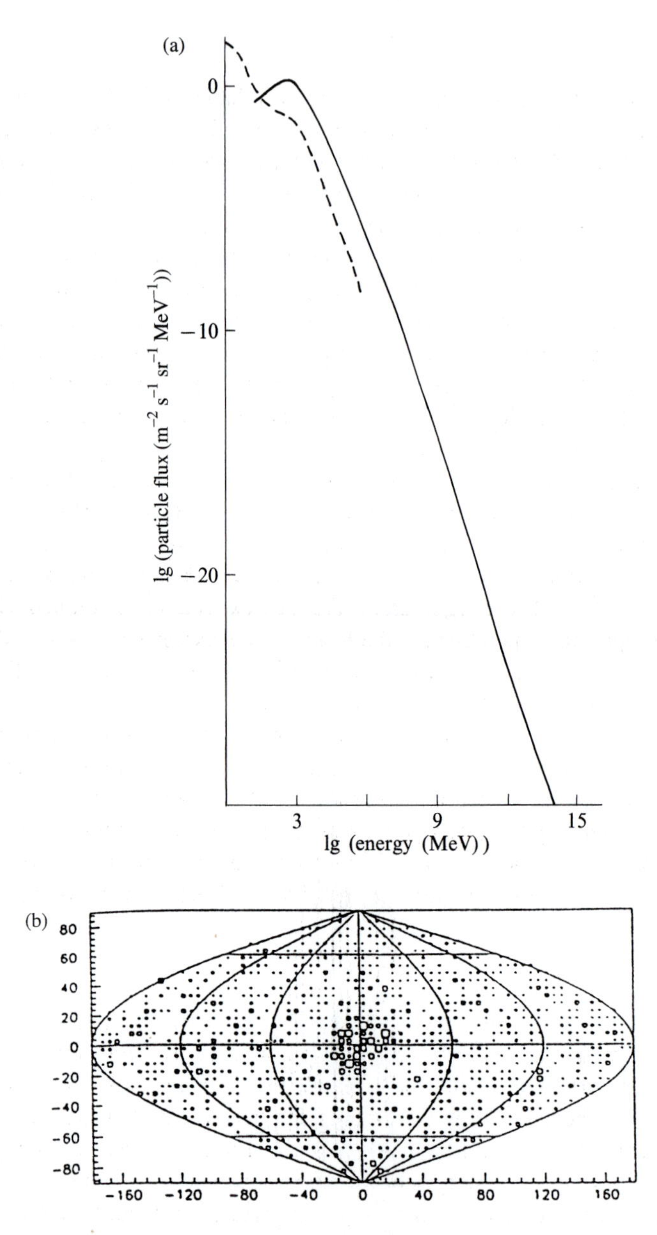

Fig. 1.8 (a) The flux of cosmic rays of different kinetic energies reaching the top of the Earth's atmosphere. Solid line: protons and nuclei. Broken line: electrons. (b) The image of the Sun taken by means of neutrinos in a celestial coordinate system in which the Sun always sits at the centre of the plot. The data are divided into $4^\circ \times 4^\circ$ bins and a box is plotted, of which the size is weighted by the number of events in each bin.

Neutrinos

These massless, chargeless particles were first postulated by Pauli and Fermi to explain an energy imbalance in β-radioactivity. Like photons, they move at the speed of light. They are exceedingly hard to detect, since the probability of their taking part in any nuclear reaction under terrestrial conditions is very low (they have a very low 'cross-section' for collisions with matter).

There are three types of neutrino, corresponding to the three known leptons: the electron neutrino, ν_e, the muon neutrino, ν_μ, and the tau neutrino, ν_τ. There is predicted to be a cosmological background of all three neutrino types left over from the hot big bang era, but it is unlikely to be detectable in the foreseeable future. Electron neutrinos have been detected from the Sun and SN1987A, muon neutrinos have been detected in terrestrial accelerators and reactors, but the tau neutrino has not been detected to date.

The world's first neutrino telescope was created by Ray Davis in South Dakota, USA. It consists of a vast tank of percloroethylene (C_2Cl_4), at the bottom of the Homestake Gold Mine to avoid contamination by cosmic rays. Any neutrinos passing through have a small chance of converting a chlorine atom ^{37}Cl to an argon atom ^{37}Ar. These argon atoms are then extracted and counted. Other neutrino telescopes now operating are the Kamiokande experiment, Japan, which uses a tank of very pure water as a Cerenkov detector, and the SAGE and GALLEX experiments, which use gallium as a detector.

The main source of neutrinos at the Earth is expected to be nuclear reactions at the centre of the Sun. Figure 1.8(b) shows an image of the sky in neutrinos made by the Kamiokande experiment, demonstrating that the Sun is the source. The rate at which all the current experiments detect neutrinos is a factor of at least two lower than is predicted by standard theoretical models of the Sun's interior. A possible explanation is that the electron neutrino is unstable and has a small non-zero mass.

Dark matter

The rotation curves of spiral galaxies, and the dynamics of binary galaxies, groups, and clusters, show that the halos of galaxies are dominated by some form of dark matter. Although part of this could be in the form of brown dwarfs or low-mass stars, the bulk is believed to be in the form of non-baryonic dark matter, probably weakly interacting massive particles (WIMPs). Such matter is needed to explain how galaxies form from the very smooth (to one part in 10^5) baryonic matter distribution deduced from the isotropic microwave background radiation at redshift $\approx 10^3$. It is also needed to account for the high value for the mean density of the universe, measured using dynamical methods from IRAS galaxy surveys.

Several experiments are now under way to try to detect the non-baryonic matter in the halo of our Galaxy. They rely on detection of the occasional nuclear recoil expected due to impact by dark matter particles, and are located deep underground to eliminate the cosmic ray background.

Gravitational radiation

According to the general theory of relativity, and other similar theories of gravitation, gravitational waves should exist. The final stages of gravitational collapse of massive stars and the coalescence of close binary star systems provide the most likely sources of gravitational radiation. J. Weber made the first attempt to detect such radiation by looking for vibrations in a large, freely suspended aluminium bar, but no believable events have been detected in this or subsequent experiments to date. The existence of gravitational radiation has, however, been inferred from the orbital changes of the *binary pulsar* (Section 4.4), a system consisting of two neutron stars in a close orbit about each other. Direct detection of gravitational waves from neutron star coalescence is confidently expected in the next generation of experiments.

1.4 Radiation mechanisms

We now look at the main ways in which electromagnetic radiation is produced in cosmic sources. We first define the *monochromatic luminosity* $P(\nu)$ of a source at frequency ν as the energy it emits per second per unit solid angle per unit frequency (units: W sr^{-1} Hz^{-1}). This means that the energy emitted per second by the source per steradian between frequencies ν, $\nu + \mathrm{d}\nu$ is $P(\nu)\,\mathrm{d}\nu$, and the total ('bolometric') luminosity can be obtained by integrating over all frequencies:

$$P = \int_0^\infty P(\nu)\mathrm{d}\nu\,.$$

The function $P(\nu)$, as a function of ν, is the *spectrum* or *energy distribution* of the source, and is an important clue to the radiation mechanism.

Blackbody radiation

If a piece of matter completely absorbs all the radiation falling upon it or, conversely, behaves as a perfect radiator when heated, then the matter radiates as a black body, and the radiation has the characteristic Planck blackbody spectrum: an element of the surface radiates with intensity

$$I(\nu) = B_\nu(T) = 2h\nu^3/c^2\{\exp(h\nu/kT) - 1\} \quad (\mathrm{W\ m^{-2}\ Hz^{-1}\ sr^{-1}}), \qquad (1.2)$$

where h is Planck's constant, k is the Boltzmann constant, and T is the temperature. For $h\nu \ll kT$,

$$\exp(h\nu/kT) - 1 = h\nu/kT$$

so $B_\nu(T) \propto \nu^2$. This is called the Rayleigh–Jeans part of the spectrum. For $h\nu \gg kT$,

$$B_\nu(T) \propto \nu^3 \exp(-h\nu/kT)\,,$$

the Wien distribution.

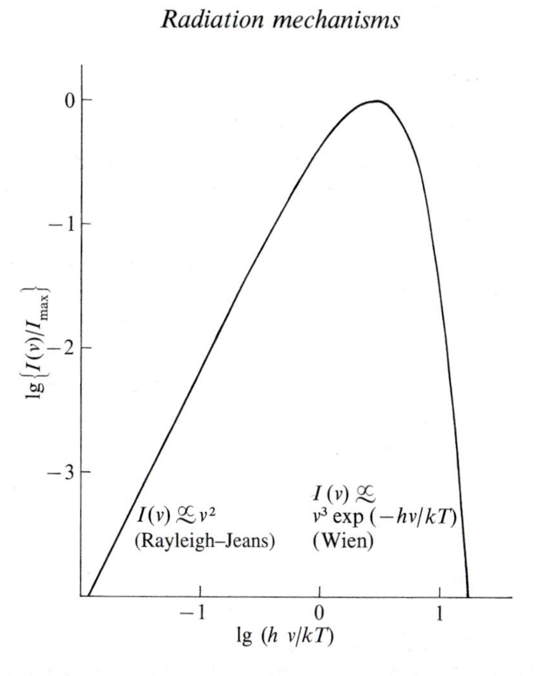

Fig. 1.9 The Planck blackbody spectrum. The intensity is given in units of the peak intensity. $I_{max} = 1.9 \times 10^{-19}\, T^3\, \mathrm{Wm^{-2}\, sr^{-1}\, Hz^{-1}}$.

The spectrum is shown in Fig. 1.9: the peak occurs at frequency $\nu = 0.354\, kT/h$. Now kT is a measure of the average thermal energy of a particle of matter (atom or molecule) and, from quantum theory, $h\nu$ is the energy of a photon of frequency ν. Thus the typical energy of the photons is of the same order as the mean kinetic energy of the matter. Blackbody radiation is the signature of matter in thermal equilibrium with radiation, with the energy shared equally between the matter and the radiation.

For the Sun, the temperature of the blackbody curve which gives the best fit to its spectrum is 5800 K (the Sun is not a perfect black body, though): most stars have blackbody temperatures in the range 2000–100 000 K. At ultraviolet and X-ray wavelengths we see thermal emission from hot gas at temperatures from 1–100 million K. At infrared wavelengths we see thermal emission from dust grains at temperatures from 10–1000 K. In the latter case, the dust grains are not perfectly efficient radiators, though, and the intensity becomes

$$I(\nu) = Q_\nu B_\nu(T)$$

where Q_ν (≤ 1) is the efficiency with which the grains radiate.

Line radiation and absorption

When an electron in an atom makes a transition from an energy level E_1 to a lower level E_2, a photon is emitted with frequency ν, where $h\nu = E_1 - E_2$.

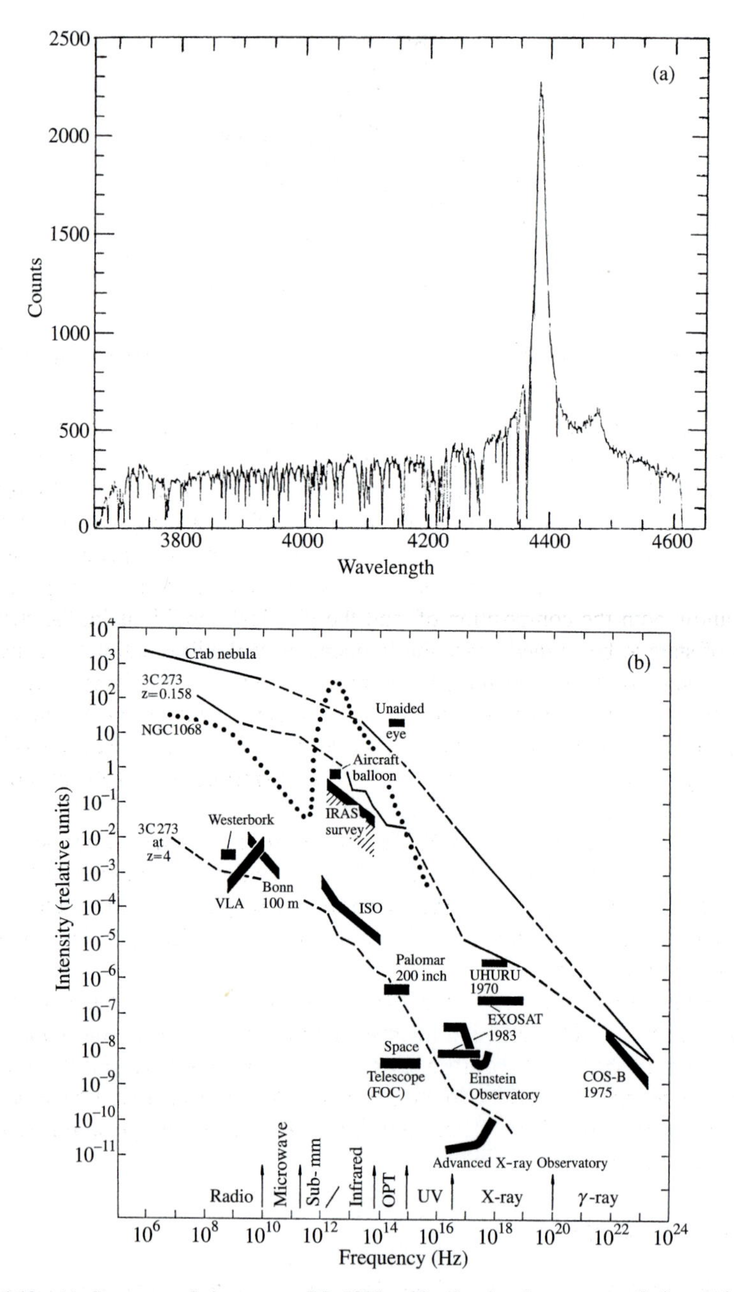

Fig. 1.10 (a) Spectrum of the quasar B2 1215 + 33, showing Lyman α emission and many absorption lines due to gas clouds between us and the quasar. Spectrum obtained by J. Bechtold using the Multiple Mirror Telescope, Arizona. (b) Continuum spectra of the Crab nebula, the Seyfert galaxy NGC1068 and the quasar 3C273 (shown at its observed redshift of 0.158 and as it would appear at $z = 4$), compared with the sensitivities of ground-based and space-borne telescopes.

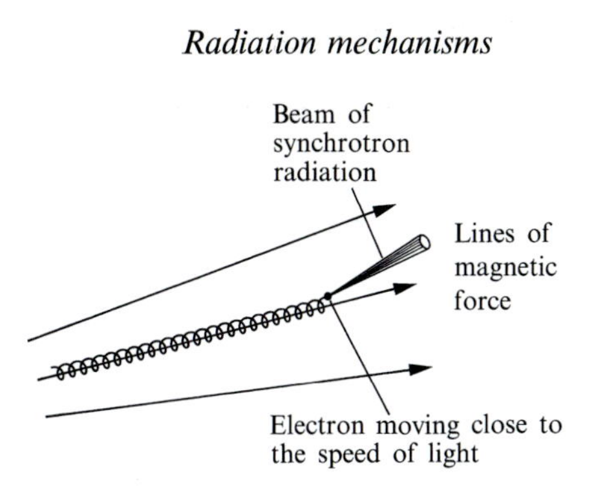

Fig. 1.11 A beam of synchrotron radiation from a relativistic electron spiralling in a magnetic field.

Similarly an atom may absorb a photon of energy $h\nu$ when an electron makes a transition from E_2 to the higher energy level E_1. These processes are very important in the surface layers of stars and result in clear emission and absorption spikes in the energy distribution $P(\nu)$ (see Fig. 1.10). If the light from a star is viewed through a prism and a narrow slit, bright or dark lines appear across the spectrum. Such lines were first noticed by Fraunhofer in the Sun's spectrum. They allow both the composition of, and the physical conditions in, the surface layers of stars to be studied, since the frequencies of the lines allow the emitting or absorbing atoms to be identified, and relative strengths of different lines give information on temperature and the number of atoms involved. For example, the transitions from higher energy levels down to the $n = 1, 2, 3, 4$ energy levels in the hydrogen atom result in the emission of the characteristic wavelengths of the Lyman, Balmer, Paschen, and Brackett series.

One of the most important spectral lines for astronomy is the 21-cm radio line of neutral atomic hydrogen, resulting from the transition from alignment to non-alignment, and vice versa, of the electron and proton magnetic axes. The distribution of neutral hydrogen in the Milky Way and other galaxies has been mapped using this line.

An important development of the 1970s was the discovery of microwave and submillimetre line radiation from interstellar molecules, arising from transitions between different states of rotational energy of the molecules. The most abundant of these molecules (apart from hydrogen, which has no rotational transitions) is carbon monoxide, which has been used to map the molecular gas in our Galaxy and other galaxies.

Synchrotron radiation

The radio emission from the Milky Way comes from cosmic-ray electrons spiralling in the Galaxy's magnetic field (Fig. 1.11). The process was first observed in particle accelerators known as synchrotrons, hence the name. It is believed to be the mechanism operating in the majority of cosmic radio sources. If the relativistic (i.e. moving close to the speed of light) electrons have

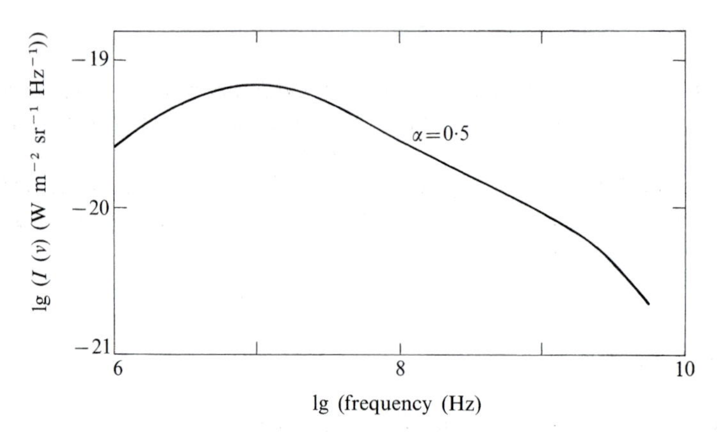

Fig. 1.12 The radio spectrum of the Milky Way. Over a wide range of frequencies it shows the power-law behaviour expected if the emission is synchrotron radiation due to cosmic-ray electrons (Fig. 1.8(a)) spiralling in our Galaxy's magnetic field.

a power-law distribution of energies (as we know they do in the vicinity of the Earth—see Fig. 1.8), then the spectral energy distribution of the synchrotron radiation will also be a power law (Fig. 1.12):

$$P_s(\nu) = A(KB^{1+\alpha})\nu^{-\alpha}\,, \tag{1.3}$$

where B is the magnetic field intensity, K is the total energy in relativistic electrons, α is the spectral index and A is a constant.

Synchrotron radiation is one of the most important examples of *non-thermal radiation*. In our Galaxy and other spiral galaxies, the relativistic electrons are believed to be accelerated in pulsars and supernova remnants. In active galactic nuclei, processes close to the event horizon of a massive black hole may be important for accelerating the relativistic electrons.

Inverse Compton radiation

This is the inverse process to Compton scattering of light by free electrons. Relativistic electrons transfer some of their energy by collisions to photons in a radiation field, the photons emerging at a higher frequency (and energy). For example, radio photons might be boosted to become X-ray photons. The main requirement is that the initial radiation field have a high intensity, so this process is likely to be important in compact sources like quasars and galactic nuclei. In fact a relativistic electron will lose its energy by synchrotron radiation in a magnetic field or by inverse Compton radiation in a radiation field, when both are present, in the proportion of the respective energy densities of the magnetic and radiation fields.

Bremsstrahlung or free–free radiation

Another process which is important for radio and X-ray sources occurs in a hot ionized gas, i.e. a gas in which outer electrons have been stripped off most atoms, leaving them as positively charged *ions*. When an electron is accelerated in the electrostatic field of an ion during a collision, it radiates energy as bremsstrahlung radiation. The total energy emitted per second by a volume V (in m$^{3)}$ of gas at temperature $T(> 10^6$ K) is approximately

$$P = 2.4 \times 10^{-28}\, T^{1/2} n_e^2\, V\ (W)\,, \tag{1.4}$$

where n_e is the number of electrons per m^3 of the gas.

1.5 Observing techniques at different frequencies

Fortunately for life on Earth, not all the radiations entering the top of the atmosphere reach the ground. Water vapour and other molecules absorb most radiation with wavelengths between 1 μm (10^{-6} m) and 2 mm, although there are a few 'windows' (narrow ranges of wavelengths through which an appreciable part of the light is transmitted by the atmosphere). The atmosphere also transmits almost no radiation with wavelengths less than 3000 Å (0.3 μm). Anyway, far ultraviolet and soft (i.e. low-frequency) X-radiation can travel only comparatively short distances through the interstellar material (gas and dust) in our Galaxy, so even from outside the atmosphere we can see very distant objects at these wavelengths only in certain directions. Low-frequency radio waves are totally reflected by the ionosphere (the same effect that makes intercontinental long-wave radio communications possible).

The percentage of light from outside our Galaxy reaching the surface of the Earth is shown in Fig. 1.13 as a function of frequency. Even from a satellite we would not be able to receive some frequencies from outside our Galaxy. In

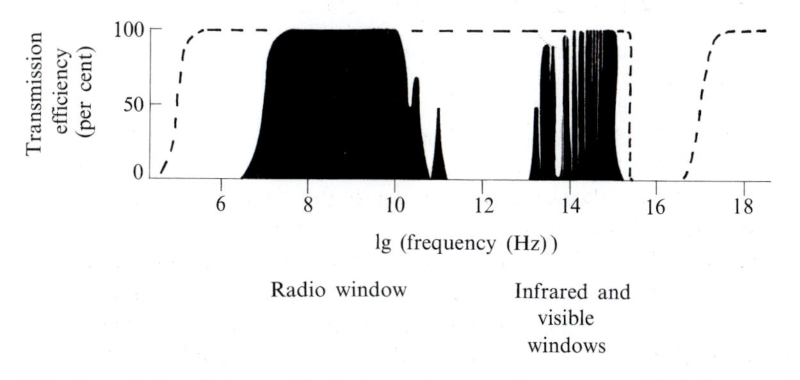

Fig. 1.13 Transmission efficiency of the Earth's environment: the percentage of light from outside the Galaxy reaching the surface of the Earth (solid curve) and a satellite outside the Earth's atmosphere (broken curve) as a function of frequency.

addition to those mentioned above, our Galaxy is opaque to the very lowest energy radio waves and the very highest energy γ-rays.

With difficulty, sources radiating at wavelengths near 1 mm, 800 μm, 400 μm, 20 μm, and 10 μm can be detected from high-altitude observatories. Balloons, aircraft and satellites have contributed to opening up these and other windows in the infrared and submillimetre region. Rockets and satellites have made ultraviolet and X-ray astronomy possible. Table 1.2 (p. 8) summarizes the observing platforms and detectors for different ranges of frequency.

1.6 The brightest sources

The *flux density* of radiation (energy per second per unit bandwidth per unit area normal to the direction of propagation) that we measure from a source of monochromatic luminosity $P(\nu)$ at distance r is given by the inverse-square law

$$S(\nu) = P(\nu)/r^2 \quad (\mathrm{W\ m^{-2}\ Hz^{-1}}), \tag{1.5}$$

provided cosmological effects (e.g. the expansion of the universe), absorption, etc. can be neglected. Thus the brightest sources we observe may be either weaker emitters that are very nearby, or distant objects that are very luminous. In Table 1.2 the brightest sources in the different frequency bands are shown.

1.7 Source counts

Suppose that at frequency ν we catalogue all the sources in the sky as a function of the flux density $S(\nu)$. Let $N_\nu(S)$ be the number of sources per steradian having flux density greater than S at frequency ν. We shall see later that this function provides an important test of cosmological models. At this point we merely note the faintest flux densities reached by present-day techniques in the different frequency bands, the corresponding value of $N_\nu(S)$, and the main contributing types of source (Table 1.2).

1.8 Integrated background radiation

If we point a telescope at a region of the sky free of bright sources, we can measure the total integrated background flux density from all the sources in the sky at all distances. This will depend on the size of the telescope beam, so we naturally measure this background in terms of the *intensity* of the radiation, the flux density per unit solid angle, $I(\nu)$ W m^{-2} Hz^{-1} sr^{-1}. The current state of observations of this quantity is shown in Fig. 1.14. Only at radio and X-ray frequencies have we succeeded in detecting a background that probably does come from discrete sources. The most striking feature of Fig. 1.14 is the 2.7 K blackbody background radiation at centimetre and millimetre wavelengths. This is believed to be the relic of the 'fireball' phase of the big-bang universe. The optical and infrared region is unfortunately dominated by atmospheric emission

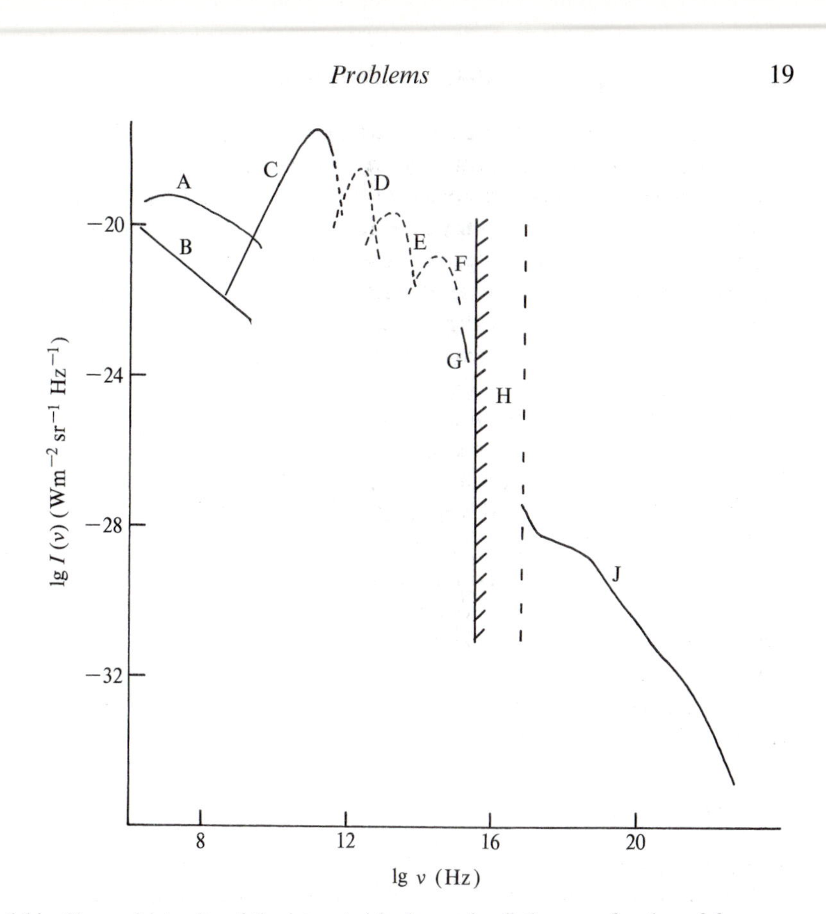

Fig. 1.14 Observed intensity of the integrated background radiation as a function of frequency (broken sections indicating theoretical predictions). A: background from the Milky Way (Fig. 1.12); B: contribution of extragalactic sources with steep radio spectra; C: 2.7 K blackbody cosmic microwave background; D: background from infrared galaxies; E: foreground emission from interstellar and interplanetary dust (at Galactic and ecliptic poles, respectively); F: combined effect of zodiacal light and scattered star light; G: ultraviolet background at Galactic pole; H: unobservable extreme ultraviolet (except perhaps at high frequency end); J: the X-ray and γ-ray background at high Galactic latitudes.

(the airglow), scattered light from interplanetary dust (zodiacal light), light from the stars of the Milky Way and emission from interplanetary and interstellar dust.

1.9 Problems

1.1 The microwave background radiation has the spectrum of a black body of temperature 2.7 K. Show that the total integrated intensity of a black body of temperature T:

$$I = \int I(\nu)\,\mathrm{d}\nu = \int B_\nu(T)\,\mathrm{d}\nu$$

satisfies $I = AT^4$ (you do not need to evaluate A: it is in fact 1.805×10^{-8} W m^{-2} sr^{-1} K^{-4}).

The energy density u, of a radiation field, is given by $u = 4\pi I/c$. Evaluate I and u for the microwave background radiation. For what value of T would u equal the energy density of an atomic nucleus? (You may estimate this using $m_H = 1.67 \times 10^{-27}$ kg, $c = 2.998 \times 10^8$ m s^{-1}, radius of hydrogen atom $= 10^{-15}$ m.)

1.2 The intensity of the radio emission from our Galaxy at high Galactic latitudes is approximately

$$I(\nu) = 2.5 \times 10^{-20} \, (\nu/10^8)^{-0.5} \ \mathrm{Wm^{-2}\,Hz^{-1}}$$

from $\nu = 10^7 - 10^{10}$ Hz.

Evaluate the total integrated intensity $I = \int I(\nu)\, d\nu$, and give the ratio of this to the total intensity of the microwave background radiation from question 1.

2
Our Galaxy and other galaxies

2.1 Introduction

In this chapter we look at the structure and evolution of our Galaxy and other galaxies, and introduce the different kind of 'active' galaxy—starburst galaxies, radio galaxies, Seyfert galaxies, and quasars.

The main constituents of galaxies are stars and interstellar gas and dust, and the main evolution of an isolated galaxy consists of stars condensing out of clouds of gas and dust, undergoing thermonuclear reactions, losing mass in winds from their surface, and finally either cooling down or dispersing much of their material in spectacular explosions.

By tracing out the distribution of populations of stars of different ages in our Galaxy, we can get a good idea of the way that the Galaxy has evolved with time and what the last stages of its formation were like. Other types of galaxy can be explained as similar in structure to our own, and of the same age, but with differing rates of star formation. Interactions and mergers between pairs of galaxies also play a significant part in the evolution of some, perhaps most, galaxies.

In their radio, infrared, and X-ray properties, galaxies show an enormous range of power and size. The different types of 'active' galaxy can be understood in terms of massive bursts of star formation and/or massive black holes in galactic nuclei. Our viewing angle also plays a part in explaining some of the range of phenomena observed.

2.2 Star formation and the interstellar medium

The Orion nebula (Fig. 2.1) provides a good example of a gas cloud out of which stars are currently forming. The Trapezium stars have condensed very recently, and one of them is heating up and exciting part of the gas to give the visible cloud of ionized hydrogen or H_{II} region, as it is called. Nearby we can see cooler condensations, radiating mainly in the infrared, which are probably even younger 'protostars'. The region covered by the photograph is just a small part of a giant complex of molecular clouds, 300 light years across.

The gas and dust between the stars, or *interstellar medium*, is found in several distinct phases. Much of it is in the form of diffuse clouds of neutral atomic hydrogen (H_I) or dense clouds of molecular hydrogen (H_2). Between these clouds is a sparser and very much hotter phase of ionized gas, mainly the relics of supernova explosions. In all of these phases there is an admixture of about 24 per

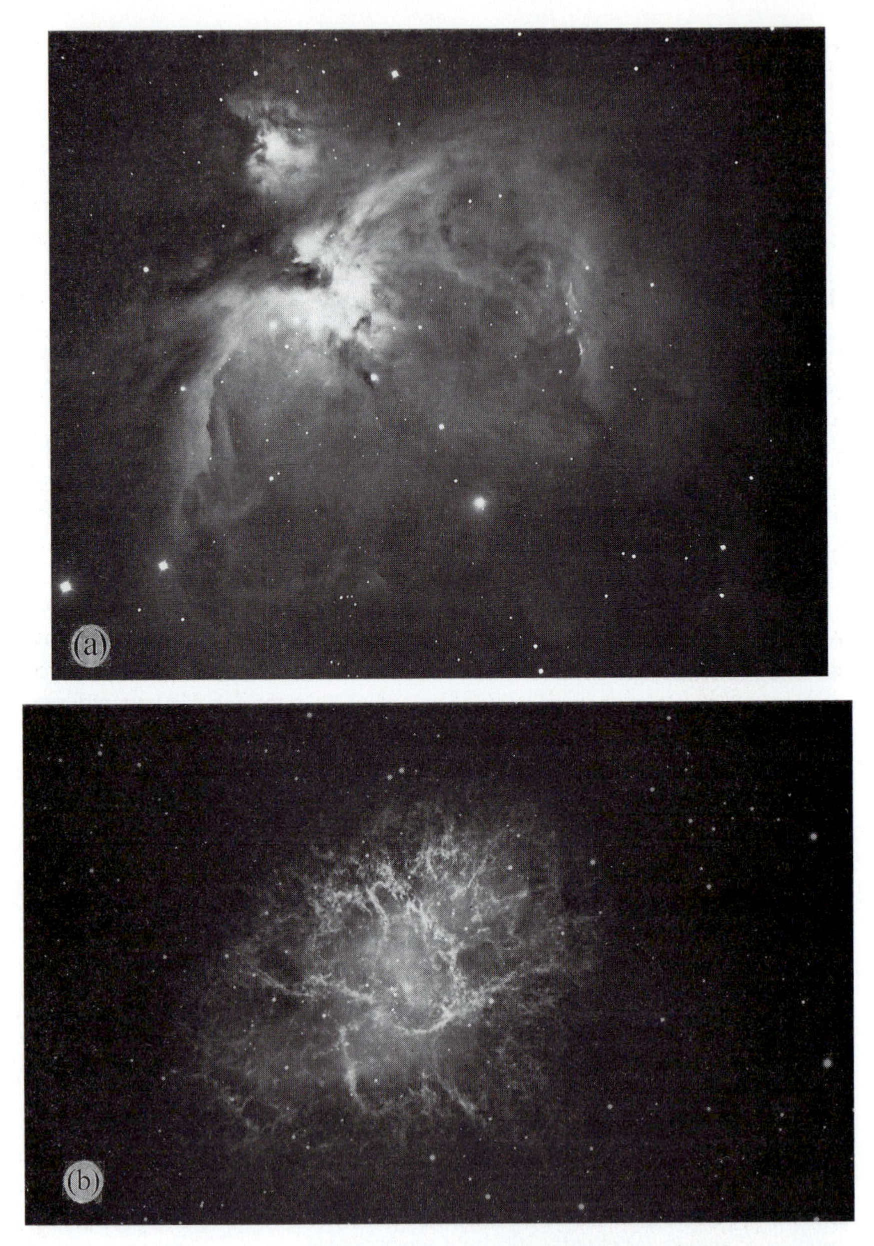

Fig. 2.1 (a) The birth of a star: the Orion nebula, part of a gas cloud out of which new stars are forming. (b) The death of a star: the Crab nebula, the relics of a star which exploded as a supernova in AD 1054. The core of the star contracted to form a neutron star, which can be seen as a pulsating radio, optical, and X-ray and γ-ray source. Photographs from Mt Wilson (a) and Mt Palomar (b) Observatories (copyright by the California Institute of Technology and the Carnegie Institute, Washington).

cent helium by mass, and about 2 per cent of heavier elements, mostly in the form of small grains of carbon or silicate dust.

The formation of a new star starts when some portion of a dense cloud of molecules and dust has a slightly higher density than average for the cloud. The self-gravitation of this region tends to make it fall together and, as this happens, what is by now the protostar heats up. The infalling material tends to form a thick disc or torus around the protostar and a bipolar outflow along the axes of this disc is often seen from newly forming stars. Eventually the central temperature of the condensation becomes high enough for nuclear reactions to start. For a cloud composed mainly of hydrogen this happens when the temperature gets above about 10^7 K: hydrogen then fuses to form helium. The pressure in the protostar builds up until the pressure gradient can balance gravity, and the central and surface temperatures then adjust themselves so that the amount of heat radiated at the surface is balanced by the amount of energy generated in the nuclear reactions at the centre. A star is born.

2.3 The evolution of a star

While a star is 'burning' hydrogen it turns out that the luminosity L and surface temperature T of the star both depend only on the mass M of the star (and very slightly on the amounts of elements other than hydrogen that are present):

$$L \underset{\sim}{\propto} M^{3.4}; \quad T \underset{\sim}{\propto} M^{0.5}. \tag{2.1}$$

Thus stars of different mass lie on a line (the *main sequence*) in a luminosity–temperature, or *Hertzsprung–Russell* (HR), diagram. Since colour is a good indicator of temperature, this is usually used as the horizontal axis by astronomers. Another good indicator of surface temperature is the type of spectral lines found in the spectrum of the star, and stars can be classified according to their *spectral type*. The spectrum of a hot star is dominated by emission lines of hydrogen and helium. In a star of medium temperature, like the Sun, metallic absorption lines dominate. And in a cool star we see absorption by molecular bands. A schematical HR diagram is shown in Fig. 2.2, with the main sequence and the track of a forming star indicated.

When most of the hydrogen in the hot central core has been fused into helium, the star undergoes a dramatic change. The surface of the star becomes cooler and redder, and the star grows in size by a large factor, becoming a *red giant*. The core meanwhile contracts and becomes hotter until, if the star is massive enough, helium starts to fuse into carbon—the *helium flash*. For the most massive stars subsequent thermonuclear reactions progressively fuse carbon into nitrogen, nitrogen into oxygen, and so on, through elements of higher and higher atomic number, up to iron. For elements with atomic number higher than iron the fusion reaction is *endothermic*, i.e. it absorbs energy instead of giving it out, so once the core is composed of iron it can no longer remain in equilibrium. The core of

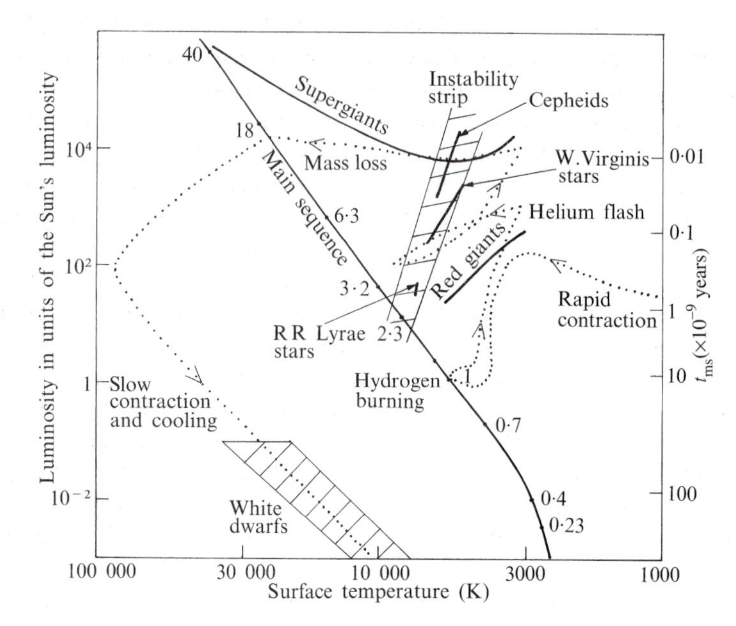

Fig. 2.2 The Hertzsprung–Russell diagram, which shows how the luminosity and surface temperature of stars change with time. The dotted curve is the predicted evolution of the Sun. The right-hand scale gives the lifetime of stars of different mass on the main sequence (points on the main sequence have been labelled with the appropriate mass in solar units).

the star collapses to form a neutron star or black hole, and the surface layers of the star are expelled in a Type II supernova explosion.

Elements with atomic number higher than iron are formed by a different process—*neutron capture*. The rare-earth elements are formed by the slow irradiation of iron nuclei by neutrons, in red-giant stars, and the radioactive elements are formed by rapid neutron capture in supernova explosions.

2.4 Final stages: white dwarfs, neutron stars, and black holes

The final stages of a star's life depend on its mass, and are not yet completely understood. If the mass is much greater than that of the Sun, the star will explode violently as a supernova, showering the interstellar gas with a mixture of all the elements that have been produced in its nuclear reactions. The most spectacular example of a relic of such an explosion is the Crab nebula (Fig. 2.1(b)), the remains of a supernova explosion observed by Chinese astronomers in AD 1054.

Less spectacular variations and ejections also take place during the later stages of stellar evolution, e.g. the Cepheid and RR Lyrae variable phases (short-period pulsations that occur when a star crosses the 'instability strip' in the HR diagram—see Fig. 2.2), Mira variables (red-giant stars undergoing irregular, long-period pulsations accompanied by mass loss), planetary nebulae (caused when a star throws off a spherical shell of gas while it is on the red-giant branch) and novae (outbursts less dramatic than supernovae resulting from material deposited on a white-dwarf star by a binary companion).

In the very last stage of all, whatever is left of the star after explosions or mass loss must become either (a) a white-dwarf star, in which the density is so high that the electrons are *degenerate* (i.e. crushed together until they are touching), (b) a neutron star, in which the density is even higher and the neutrons are degenerate, or (c) a black hole, where the matter of the star collapses to such a high density that light can no longer escape from it (see Section 4.4). Only stars with mass less than about twice the Sun's mass can become white dwarf or neutron stars (in more massive stars the degeneracy pressure is not sufficient to support the star against gravity), so if more massive stars than this fail to eject most of their mass in their eruptive phases, they *must* become black holes.

Before they have cooled off too much, white dwarfs are detectable at optical frequencies, like other stars. Neutron stars, which will tend to be spinning rapidly, can be seen by pulsed radio, optical, X-ray, and γ-ray emission from their rotating magnetospheres (*pulsars*). There is a prominent pulsar in the centre of the Crab nebula, almost certainly the remains of the star that exploded in the AD 1054 supernova. Neutron stars can also be seen by X-ray emission from gas heated up by falling at high speed on to the star's (solid!) surface.

The only way to detect a black hole is through some indirect effect, e.g. if it is in a binary system with a visible companion. From the period and radius of the orbit of the visible companion we can deduce the mass of the invisible object, and if this is very large deduce that it must be a black hole, especially if X-ray emission from in-falling gas indicates that it is very compact. The X-ray source Cyg X-1 is the best candidate for a binary system containing a black hole, and several other good candidates have been found (e.g. V616 Mon, LMC X–3).

2.5 The life history of our galaxy

The oldest stars in our Galaxy are 10–15 billion years old and are composed of about 76 per cent hydrogen, 24 per cent helium, and almost no other detectable elements. The simplest assumption is that our Galaxy condensed out of a large gas cloud with this composition about 10–15 billion years ago. We shall see later that this age and this composition fit in very well with the big-bang picture of the universe (Sections 3.3 and 5.3).

Structurally our Galaxy can be divided into a disc, a nucleus, and a halo (Fig. 2.3). The stars in the Galaxy can be divided into two main populations: (a) *Population I*, bright young stars found only in the disc, associated with gas, dust, and regions of star formation; (b) *Population II*, faint old halo stars, with very low metal (i.e. elements higher than helium in atomic number) abundance.

The different parts of the Galaxy are associated with different phases of its evolution.

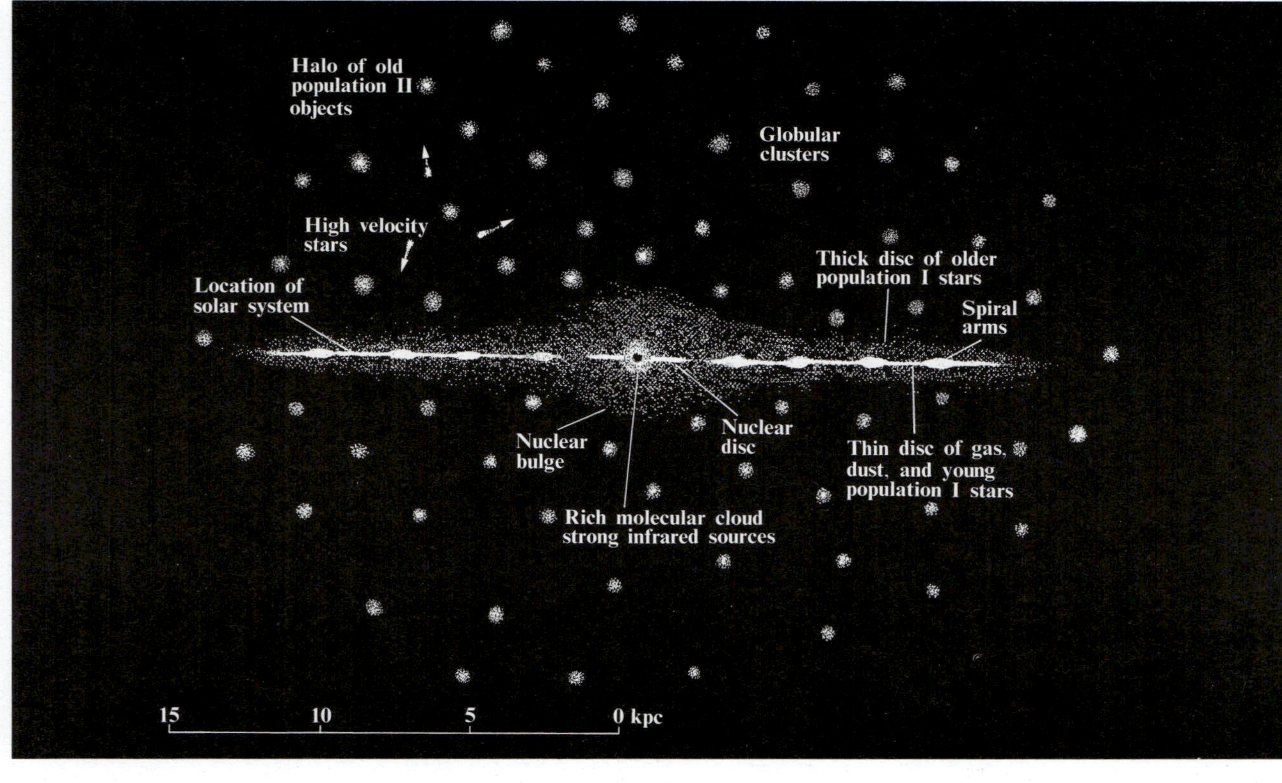

Fig. 2.3 A schematic picture of our Galaxy, seen edge-on.

The halo: phase 1

In the halo we find globular clusters containing the oldest stars in the Galaxy, and high-velocity (also old) stars. The age of these clusters can be determined by plotting their HR diagram and seeing how much of the main sequence is left. The length of time a star spends on the main sequence is approximately proportional M/L and by eqn (2.1) this means that high-mass stars have much shorter lives than low-mass stars. In a young cluster we will still find quite massive stars on the main sequence, but in an old one only the low-mass ones will still be left there, the more massive ones having already evolved away.

The age of objects in the halo shows that this was the first place where star formation occurred, and the high velocities of the stars suggests that the gas cloud out of which the Galaxy formed was in a state of rapid motion, presumably collapsing. The surface layers of halo stars are found to contain almost no other elements except hydrogen and helium (although some other elements will exist in the interiors of the more evolved halo stars). The time-scale for the collapse of the 'protogalaxy' can be shown to be about 10^8 years.

The orbital velocity of gas in the outer parts of the Galactic disc falls off with distance slower than expected on the basis of the visible matter in the Galaxy. This has led to the suggestion that our Galaxy, and other galaxies, are surrounded by a halo of dark, non-luminous matter. This could be in the form of Jupiter-sized objects, brown dwarfs (objects in the mass range 10–80 Jupiter masses which fail to ignite hydrogen-burning), massive (10^3–10^6 solar masses) black holes, or in some exotic, non-baryonic form.

The thick disc and the nucleus: phase 2

The gas that was not used up in phase 1 of the evolution of the Galaxy continued to collapse, and would have formed a concentrated core but for the effects of rotation. The protogalactic cloud must have possessed some angular momentum, since it formed a disc, with centrifugal force balancing gravity. This rotating disc was about 2000 light years thick, with a pronounced bulge towards the centre. The second generation of stars then formed, the more massive ones evolving rapidly, exploding and spraying the gas with the products of their nucleosynthesis. Thus in the surface layers of these older stars of Population I we start to see significant amounts of heavy elements.

The thin disc: phase 3

The gas still unused, with its admixture of debris from dead stars, settled down to a thin disc about 300 light years thick. Stars have continued to form in this disc of gas right up to the present day, especially in the spiral arms, which probably represent a spiral wave of star formation rotating through the disc of gas. From our position in the disc of the Galaxy we cannot see the spiral arms, but they can be traced out through 21-cm line emission from cool, neutral hydrogen (H_I) We can get an idea of how our Galaxy looks from outside from photographs of external galaxies (Figs 1.3 and 2.4).

The nuclear disc

Although we cannot see through to the centre of our Galaxy at optical wavelengths, due to obscuration by dust, we can do so in the radio and infrared. In addition to a rotating disc of gas and stars, we can see a massive cloud of molecular gas, containing strong infrared sources, probably due to hot dust grains ($10\ K < T < 1000\ K$). There is also evidence of a high-velocity (several hundred kilometres per second) outflow of gas. In many ways the nucleus of our Galaxy resembles those of the more violent objects described in Section 2.9, although on a much weaker scale. Evidence from observations of the 12.8-μm line of ionized neon suggests that the

(a)

Fig. 2.4 Examples of different galaxy types: (a) M87, an elliptical, E1. (b) NGC4565, edge-on spiral, Sb; (c) M51, face-on spiral, Sc; (d) NGC7741, barred spiral, SBa; (e) the Large Magellanic cloud, irregular. Photographs (a), (c), (e) from Lick Observatory; (b), (d) from Mt Palomar Observatory (copyright by the California Institute of Technology).

(b) & (c)

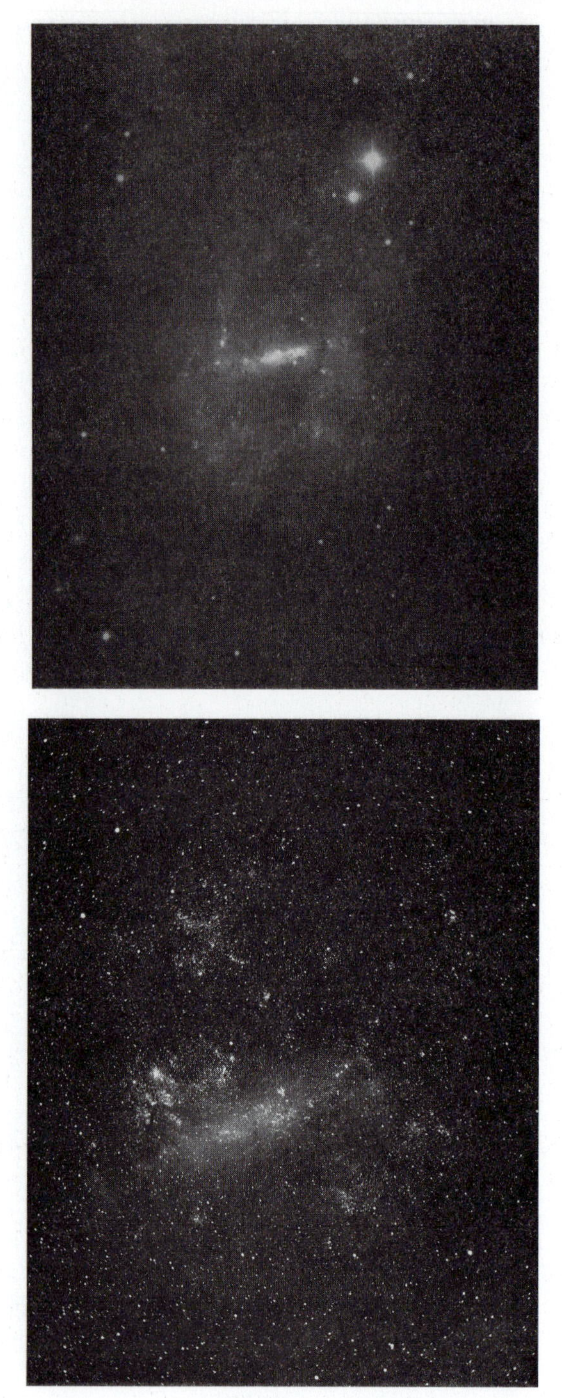

(d) & (e)

mass within the central 3 light years of our Galaxy is 8×10^6 times the mass of the Sun, and it has been speculated that this could be a black hole.

2.6 The structure and evolution of galaxies

Some examples of external galaxies are shown in Fig. 2.4. Hubble introduced a classification of galaxies, the 'tuning fork' diagram (Fig. 2.5), which is still broadly used, although it is no longer thought to be an evolutionary sequence. The *elliptical* galaxies are classified as En, by their degree of flattening, where $n = 10(a - b)/a$ and a and b are the major and minor axes, respectively. They are thought to be spheroidal systems, and this means that we cannot distinguish between a genuinely spherical galaxy and a flattened one seen face-on. However the high proportion of apparently spherical systems shows that we cannot just assume that all ellipticals are flattened systems seen at different orientations.

The S0, or *lenticular*, galaxies seem to be a transitional form between ellipticals and spirals. They contain a disc of stars, like spirals, but the stars are all old, as in ellipticals, and there is little sign of gas or recent star formation. *Spiral* galaxies are divided into normal spirals and barred spirals, the latter differing only in that their spiral arms start from the end of a prominent bar instead of from the nucleus itself. As we go along the sequence from Sa to Sc for normal spirals, or SBa to SBc for barred spirals, the nucleus becomes less pronounced, the arms become more open and less tightly wound, the proportion of gas increases, and the colour becomes bluer. It is also easier to see individual stars in nearby Sc galaxies, showing that there are more stars with high optical luminosity.

Ellipticals show little sign of gas or recent star formation and are far redder than spirals in colour. Irregular galaxies (Irr), which tend to be very blue and have a high proportion of gas, are placed to the right of the spiral sequence. This trend is summarized in Fig. 2.6, the percentage of mass in the form of neutral hydrogen (as measured by the 21-cm line) against colour.

The blue end of the main sequence corresponds to massive stars, which burn up their fuel very rapidly (10^6 years or so). Hence a blue galaxy must contain young stars, and star formation must still be going on, whereas a red galaxy with no gas (i.e. an elliptical) shows no sign of recent star formation and all its stars could be old. There are examples of all these types of galaxy in the Local Group, as shown in Table 1.1 (p. 4).

The Hubble sequence, however, is almost certainly *not* a sequence of galaxies of different age. Old stars are found in all these galaxy types, including the irregulars. The differing gas content, and proportion of young stars, along the sequence is more likely an indication of *differing rates of star formation*, with all galaxies being of about the same age (about 10^{10} years). Ellipticals formed stars rapidly in phase 1, so no gas was left to form a disc of Population I objects. Irregulars, at the other extreme, have been forming stars at a rather slow rate and still have plenty of gas left.

Clearly the *rotation* of a galaxy must also play a part in determining, for

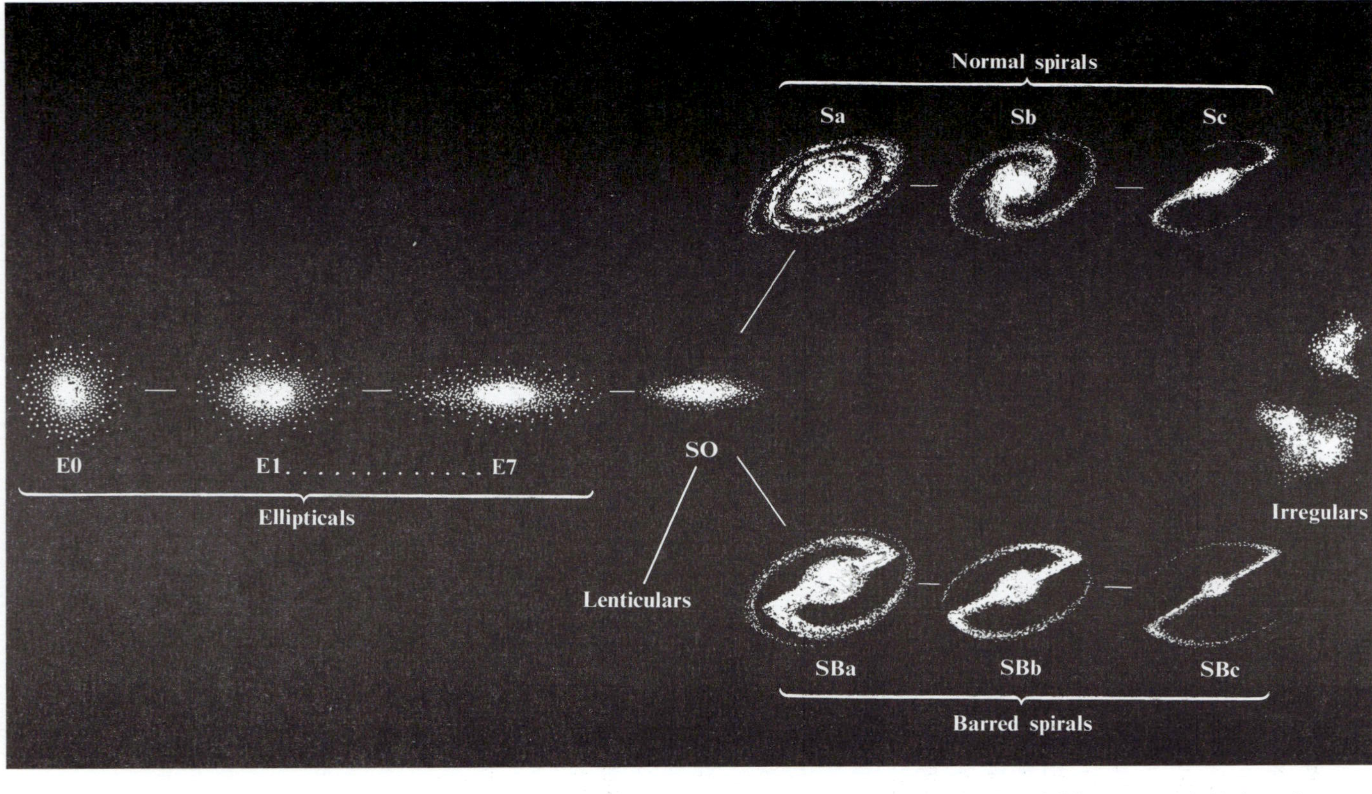

Fig. 2.5 Hubble's 'tuning fork' classification of galaxies. It is probably not an evolutionary sequence, but one of different rates of star formation.

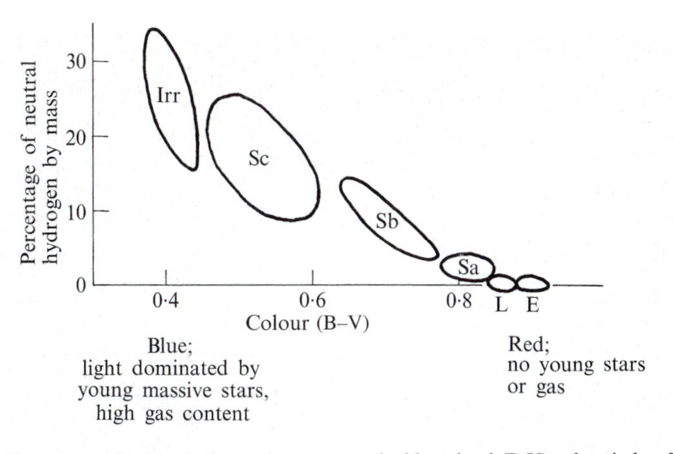

Fig. 2.6 Percentage of galaxy in form of gas, versus the blue-visual (B-V) colour index. This is the difference between the apparent magnitudes (Section 3.2) through filters with peak transmission in the blue and visual (i.e. green) parts of the spectrum. From left to right, this could be a sequence of increasing age or, more probably, increasing star formation rate.

example, the degree of flattening of an elliptical, and in the generation of spiral arms and the bars of barred spirals. But the main properties of the Hubble sequence are understood most simply in terms of a one-parameter (i.e. star-formation rate) family of galaxies of the same age.

An important property for cosmology of elliptical galaxies is that their luminosities and colours are now changing only very slowly, so they can be used as 'standard candles' (see Section 3.2).

2.7 Starburst galaxies

A major discovery of the IRAS Infrared Astronomical Satellite, launched in 1983, was that many galaxies are undergoing enormous bursts of star formation, with their main energy output occurring at infrared wavelengths. Infrared luminosities over one hundred times the entire output of our Galaxy are seen, often associated with interactions and mergers between pairs of galaxies (see Fig. 2.7). The main power of the starburst would be in the form of ultraviolet light from massive stars, but most of this is absorbed by dust and reradiated at far-infrared wavelengths. A less extreme example of a starburst can be seen in the 30 Doradus ('Tarantula' nebula) region of the Large Magellanic cloud.

It is possible that most of the star formation that has occurred in spiral and elliptical galaxies took place during starburst episodes, driven by interactions with neighbouring galaxies.

Normal spiral galaxies radiate at infrared wavelengths too, because their starlight is absorbed by interstellar dust grains and reradiated in the infrared. Typically about 30 per cent of the total power of the galaxy emerges in the far infrared (10–200 μm).

(a)

Fig. 2.7 (a) The starburst galaxy M82. (b) Radio contours superimposed on an optical photograph of the spiral galaxy M51. Radio map obtained at Westerbork.

2.8 Radio properties of galaxies

Galaxies are copious emitters of radio waves, by both the bremsstrahlung and synchrotron radiation mechanisms (Section 1.5). Only in a few cases, however, does the radio power amount to more than 1 per cent of the total power output of the galaxy.

Spirals

The radiation from spiral galaxies is concentrated in the disc of the galaxy, with a total power in the range 10^{29}–10^{33} W. Along the spiral arms we see a chain of H_{II} regions, clouds of hot (10^4 K) gas ionized by newly formed massive stars. These H_{II} regions emit thermal bremsstrahlung radiation in the radio band, recognizable by its characteristic flat spectrum ($\propto \nu^{0.1}$). We see also synchrotron radiation from relativistic electrons accelerated in pulsars and supernova remnants, and spiralling through the galaxy's magnetic field. This emission is concentrated along the spiral arms (Fig. 2.8). In addition there are weaker sources in the nuclei of spirals, with a

(b)

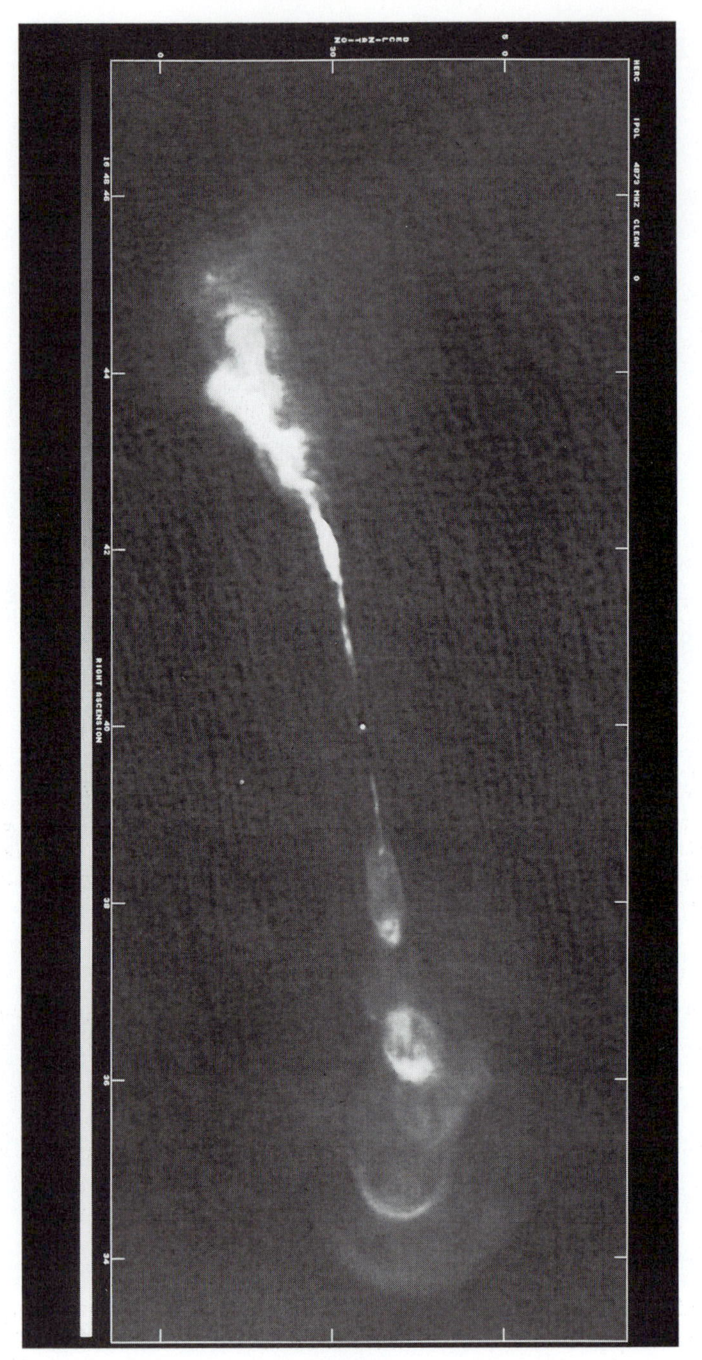
HERC IPOL 4873 MHZ CLEAN 0
RIGHT ASCENSION
16 48 46
44
42
40
38
36
34
30
0

(a)

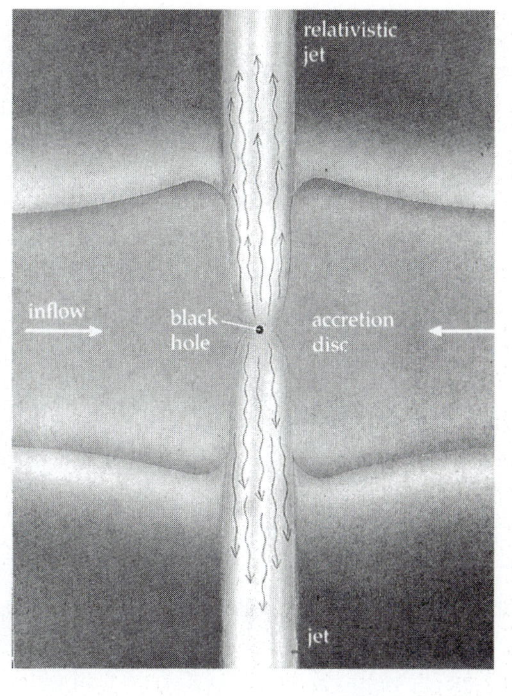

(b)

Fig. 2.8 (a) Radio map of the powerful radio galaxy, Her A, showing the characteristic double jet structure. The small dot in the centre indicates the location of the optical galaxy, where a compact radio source is centred. Map obtained at VLA, New Mexico. (b) Black hole model for the origin of double-beam radio sources. Material falls into an accretion disc surrounding the black hole and is funnelled towards the black hole, where it is accelerated to relativistic speeds along the axis of the black hole.

typical size of 1000 light years and a power of up to 10^{31} W.

There is an excellent correlation between the radio and far-infrared luminosities of spiral galaxies of the form

$$P(1.4\,\mathrm{GHz}) = 0.01\,P(60\,\mu m) \tag{2.2}$$

which arises because the radiation in both wavebands is an indirect consequence of star formation.

Ellipticals

The radio emission from elliptical galaxies usually comes from two components symmetrically placed on either side of the galaxy, at total separations of 10^4–10^7 light years from the galaxy (see Fig. 2.8). Total powers range from 10^{33} to 10^{39} W and the minimum energies required in the magnetic field and relativistic

Fig. 2.9 The nearby powerful radio galaxy Cen A (NGC5128), an elliptical with unusual dust lanes across it. Photograph from Mt Palomar Observatory (copyright by the California Institute of Technology and the Carnegie Institute, Washington).

particles range up to 10^{53} J, posing severe theoretical problems. Models not only have to explain this huge energy, which is usually assumed to arise in the galactic nucleus, but also the strange, often impressively linear, double structure. A collimated double beam of relativistic particles is believed to be powered by material (gas or stars) falling into a massive black hole ($\sim 10^8$ solar masses). The diffuse lobes of emission seen far from the galactic nucleus are believed to be generated by interaction of these beams with intergalactic gas and magnetic fields. The structure of the radio sources is correlated with the radio power, weaker sources tending to have a more diffuse and irregular structure. It remains a mystery why these double radio sources are seen only in elliptical galaxies.

A weaker core source is also found in the nucleus of many ellipticals, with typical powers in the range 10^{31}–10^{34} W.

Although the radio properties of galaxies show a wide range of powers and characteristics, it is roughly true that the radio power increases with the optical power (and hence with the mass of the galaxy). Approximately

$$P_{\text{radio}} \propto P^2_{\text{optical}} \tag{2.3}$$

Figure 2.9 shows the galaxy NGC5128, the most powerful radio emitter in our immediate neighbourhood. The main radio lobes lie far off the region of the galaxy shown. Along the dust lane of this peculiar elliptical many young blue stars are found, indicating a burst of star formation about 25 million years ago. It is now believed that the dust lane may be the debris of a spiral galaxy which collided and merged with the elliptical galaxy.

Rapid stellar motions in the nucleus of the powerful radio-emitting elliptical galaxy M87, in the Virgo cluster, point to the presence of a dark compact mass of several thousand million solar masses, possibly a black hole. Thus both galaxy mergers and massive black holes appear to be involved in the generation of these powerful radio galaxies.

2.9 Active galactic nuclei (AGN): Seyferts and quasars

Two types of object show evidence of violent, transient activity: Seyfert galaxies and quasars. It is hard to draw a sharp line between the different classes, since each type merges into the other. In each case we are seeing violent activity in the nuclei of galaxies.

Seyfert galaxies

These are galaxies (usually spirals) with an intense, often variable, star-like nuclear region. The spectra of these nuclei show very strong, broad emission lines, arising from hot gas rather than from stars. Colours are bluer than average, suggesting the presence of very hot gas (30 000 K) or a non-thermal component to the optical continuum. In some cases there is evidence for the violent ejection of material, at velocities of several thousand kilometres per second.

Seyfert nuclei have been found to be unusually strong infrared sources, probably due to emission by dust. The dust may be heated either by a recent burst of star formation or by the compact optical continuum source. Seyferts are only slightly more active than normal spirals at radio wavelengths.

An important subclassification of Seyferts is into *Type 1*, which have very broad wings to their emission lines, corresponding to relative motions of thousands of kilometres per second, and *Type 2*, which do not. The idea that this distinction is due to dust obscuration of the dense, hot gas responsible for the broad wings is gaining acceptance. The dust may be in the form of a torus surrounding the compact optical source, so that ultimately the distinction is mainly a matter of the direction of viewing.

Some radio galaxies show compact optical sources and strong emission line spectra without broad wings. Originally called N-galaxies, they are now generally called narrow-line radio galaxies (NLRG). They appear to be the analogy of Type 2 Seyferts in elliptical galaxies. Weak compact optical sources have been found in the nuclei of classical radio galaxies like Cygnus A, suggesting that most radio galaxies may show such an optical core at some level.

Quasars

These include radio sources associated with quasi-stellar optical objects (QSRs) and also quasi-stellar optical objects with very weak or undetected radio emission (QSOs). These two classes are also sometimes called *radio-loud* and *radio-quiet quasars*. The spectra of quasars show broad emission lines from hot gas only (there are no stellar lines) (see Fig. 1.10(a)). For many low-redshift quasars, distances can be determined by association with neighbouring clusters of galaxies at the same redshift. The optical powers range from a few times up to a thousand times that of a normal luminous elliptical galaxy, assuming their redshifts are cosmological (see Section 3.3). Their colours are usually very blue.

Quasars are closely linked with Type 1 Seyfert galaxies. In fact the distinction is probably merely one of the relative strength of the compact optical source and the surrounding galaxy. A quasar is simply a Seyfert in which the host galaxy cannot be seen on the Sky Survey photographs because of the strength of the compact optical source. Long imaging integrations on low-redshift quasars often reveal the presence of the host galaxy.

The radio sources associated with radio-loud quasars fall into two types: (a) double sources similar to those of elliptical radio galaxies; (b) compact sources, often variable on a time-scale of a month or less. It is remarkable that objects not much larger than the solar system, putting out up to a thousand times as much energy as the whole Milky Way, can vary their power so rapidly. A subclass of quasar with compact radio sources which are violently variable at optical, infrared, and radio wavelengths have been given the name *blazar*, derived from the prototype object BL Lac. It has been suggested that these are quasars where we are viewing the source from a direction closely aligned with the beam responsible for the double radio sources. This interpretation is supported by the fact that in such sources we often see apparently faster-than-light motions as the source varies and blobs of emission move out from the central regions. The easiest interpretation of such apparent motions is in terms of ejection of material at speeds close to the speed of light close to the line of sight.

In radio-quiet quasars the radio power, if detected, is similar to that in spiral galaxies and these sources satisfy the same relation between radio and far-infrared power as normal spirals (eqn (2.2)). This suggests that QSOs differ from normal spirals only in the presence of the compact optical source and strong emission lines. It is less obvious for these objects that a massive black hole has to be invoked to explain them, since the compact optical source and emission line spectrum can probably be explained in terms of a violent starburst and its associated supernovae and their remnants.

2.10 X-rays from galaxies and quasars

The brightest X-ray sources on the sky are Galactic binary star systems in which one of the stars is a white dwarf, neutron star, or black hole. X-rays are also seen

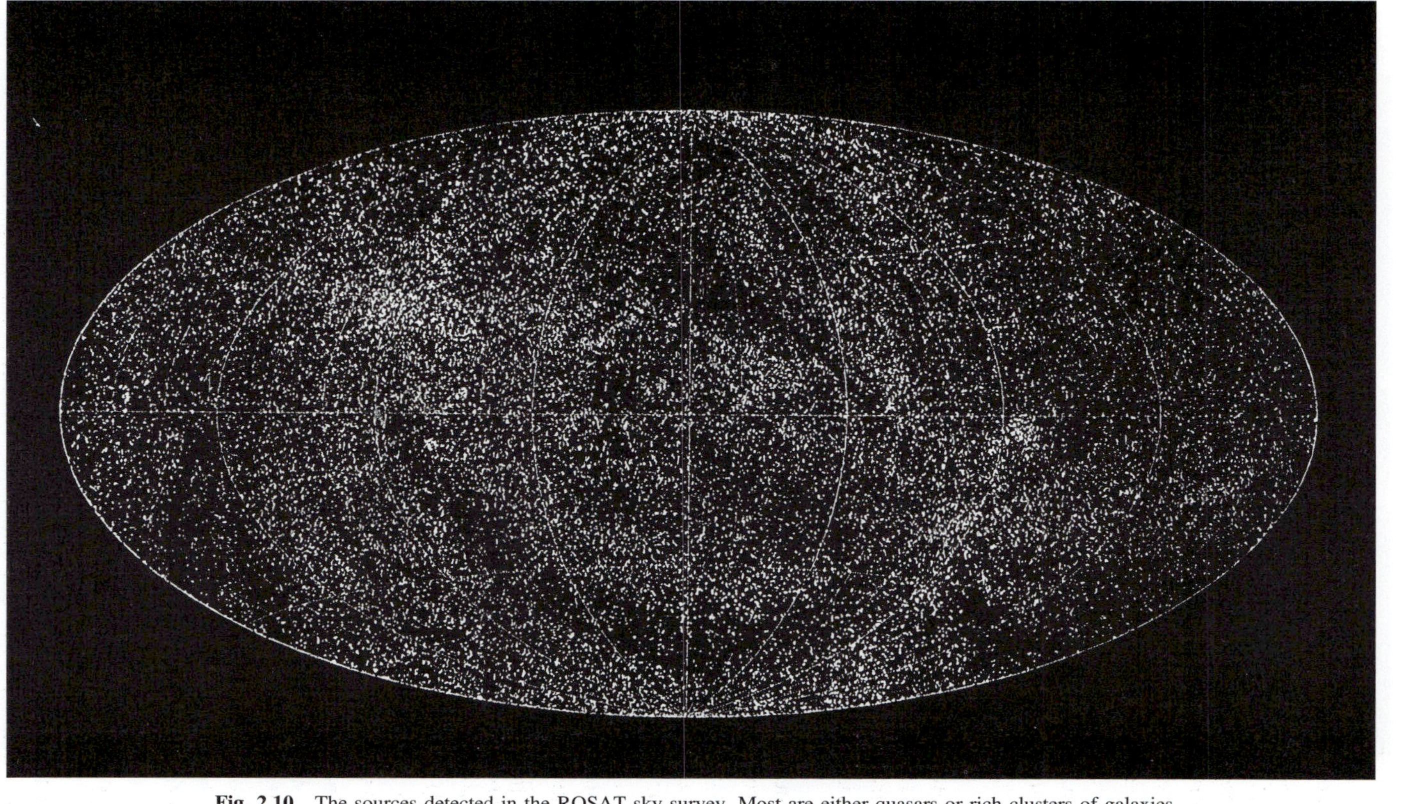

Fig. 2.10 The sources detected in the ROSAT sky survey. Most are either quasars or rich clusters of galaxies.

Fig. 2.11 The quasar 3C273 (at centre of picture).

from pulsars and supernova remnants and from the coronae of hot, luminous stars. Similar types of sources are seen in M31 and the Magellanic clouds. The total X-ray luminosity of M31 is about 2×10^{32} W.

Far more powerful X-ray sources are seen in the nuclei of active galaxies. The radio galaxies M87 and NGC5128, for example, both have nuclear X-ray luminosities of about 10^{36} W. Seyfert galaxies have X-ray luminosities in the range 10^{34}–10^{38} W, and quasars in the range 10^{36}–10^{40} W. Some of these active galaxies have their main energy output in the X-ray band. The emission mechanism is probably the synchrotron or inverse Compton process.

Tens of thousands of X-ray quasars have been found by the ROSAT survey mission (see Fig. 2.10), launched in 1990.

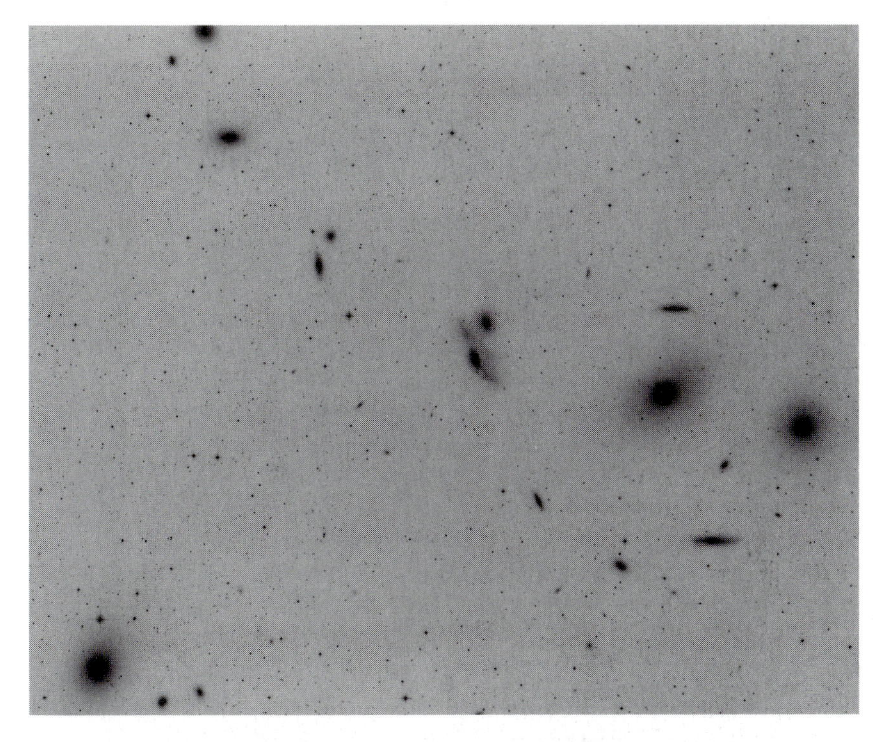

Fig. 2.12 Part of the Virgo cluster of galaxies.

So far only a few extragalactic sources, like the quasar 3C273 (Fig. 2.11), the radio galaxy NGC5128, and the Seyfert galaxy NGC4151, have been clearly identified in the γ-ray band.

2.11 Clusters of galaxies

While most galaxies occur in small groups similar to the Local Group, about 10 per cent occur in rich and spectacular clusters like the Virgo Cluster (Fig. 2.12) containing hundreds to thousands of galaxies and with total masses ranging up to 10^{15} solar masses. Cross-correlation of clusters with the general galaxy distribution, and large-scale maps of the galaxy density, show that the extent of these clusters is very large (up to a hundred million light years), and they join smoothly onto neighbouring clusters. Thus concepts like 'supercluster' are not well defined. The galaxy distribution is found to be clustered on all scales up to at least one hundred million light years.

There is also evidence for huge sheets of galaxies (one which stretches between the Coma and Hercules clusters is known as the 'Great Wall of Galaxies') and voids, on scales of 100–300 million light years. One of the questions which cosmologists face is, are these a generic aspect of the galaxy distribution? Topological studies to date are consistent with the galaxy distribution having arisen from an initially random, Gaussian distribution with a sponge-like topology (i.e. there is symmetry between regions of high and low density).

About 10 per cent of the mass of a typical rich cluster of galaxies is in the form of hot intergalactic gas at a temperature of around a hundred million Kelvins (10^8 K). This emits strongly at X-ray frequencies by the thermal bremsstrahlung mechanism (Section 1.4). A spectral line of ionized iron, Fe_{XXV}, has been seen from several of these cluster X-ray sources, from which it can be deduced that the abundance of iron in this intergalactic gas is at most a factor of three lower than in the Sun. Since iron is made in stars, it follows that much of the intergalactic gas in clusters of galaxies must have been ejected from the galaxies.

The intergalactic gas in rich clusters also makes its presence felt through 'tail' radio galaxies, in which radio-emitting plasma appears to have been swept backwards by the motion of the galaxy through the gas.

2.12 Problems

2.1 Given that the luminosity of stars is related to their mass by $L \propto M^{3.4}$ and that the total lifetime of the Sun is 10^{10} years, estimate the lifetimes of stars of mass 0.01, 0.1, 10, and 100 solar masses.

2.2 From the discussion given in Chapter 2, give a brief account of the appearance of our Galaxy (i) 10^9 years, (ii) 10^{11} years, (iii) 10^{13} years after it was formed.

3
The empirical basis for cosmological theories

3.1 Introduction

We can make progress towards answering the questions 'what is the structure of the universe?' and 'how has the universe changed with time?' only if we believe that the universe has a simple over-all structure.

In our immediate locality the universe has an exceedingly complex structure. If we had to construct a mathematical model of the human brain, and had as little observational information about it as we have about the universe, we would hardly know where to start. However, a human brain is a highly atypical region of the universe. For one thing its density is about 10^{30} times the present average density of matter in the universe.

But suppose we start looking at things on the large scale—on the scale of the distances we have been mentioning in Chapters 1 and 2. Then the Earth, and all its structure and detail that is so important to us, becomes a minute speck of dust. On this scale the universe may start to look simple. Before we can start to discuss what sort of properties the universe might have when looked at in this way, we must establish a cosmological distance scale.

3.2 The distance scale

Cosmological distance can be measured in a variety of ways, all involving the observation of light signals. Unfortunately, the most reliable methods can be used only over a short range of distances. As we look further out, we are forced to use progressively less reliable methods.

Radar techniques

These can be used to measure the distances of only our nearest neighbours in the solar system.

Parallax

The apparent change in direction of a source as the Earth goes round the Sun gives distances for the nearest few thousand stars. The distance is measured in *parsecs*; a distance is 1 parsec (pc) ($= 3.09 \times 10^{16}$ m $= 3.26$ light years) if the mean radius of the Earth's orbit subtends $1''$ of arc. Distances measured in this way give us the luminosities of nearby stars of different type. The apparent brightness of certain types of star can then be used as a distance indicator. From ground-based telescopes parallax can be reliably measured out to 30 pc. within which distance there are a few thousand stars. The Hipparcos satellite, launched in 1989, has measured parallaxes for tens of thousands of stars out to several hundred parsecs.

Moving cluster method

If the stars in a cluster of stars are moving parallel to each other, their apparent motions across the sky (or 'proper' motions) appear to converge to a point on the sky. If ϕ is the angle between this direction and that of the cluster, then

$$\tan\phi = v_t/v_r$$

where v_t is the velocity of the stars tangential to the line of sight and v_r is their radial velocity. v_r can be measured from the Doppler shift and v_t is related to the angular velocity of the stars' motion on the sky, ω, by

$$v_t = \omega d$$

where d is the distance. Hence, $d = v_r \tan\phi/\omega$. Distance up to about 200 pc can be measured in this way.

Luminosity distance

The luminosity distance d_{lum} is defined as the distance an object would appear to have if the inverse-square law held exactly (cf. eqn (1.5)):

$$d_{lum} = (P/S)^{1/2}, \tag{3.1}$$

where P is luminosity and S is the flux density of radiation. The luminosity distance provides a ladder of distance using objects ranging from moderately luminous variable stars to whole clusters of galaxies. All that is needed is a class of source (a *standard candle*) with not too great a spread in luminosity P. At visible frequencies astronomers have got into the rather odd habit of measuring flux density and luminosities in terms of *apparent magnitude m* and *absolute magnitude M*, where the magnitude scale in such that there are five magnitude steps per range of brightness of 100, with the additional peculiarity that the greater the magnitude the fainter the source. Thus

$$m = A - 2.5 \lg S \tag{3.2}$$

This scale was adopted in the nineteenth century to agree approximately with the brightness classification given by the Greek astronomer Hipparcos for bright stars in the second century BC. The constant A depends on the range of the spectrum

Table 3.1 The distance scale

Distance indicator	Luminosity distance	Diameter distance	Range
Variable stars	RR Lyrae		150 kpc
	Cepheids		20 Mpc
Bright stars	Brightest stars in galaxies		10 Mpc
Eruptive stars	Novae		20 Mpc
	Supernovae		400 Mpc
Extended objects in galaxies	Planetary rebulae	Planetary nebulae	20 Mpc
	H_{II} regions	H_{II} regions	25 Mpc
	Globular clusters	Globular clusters	50 Mpc
Galaxies	Luminosity of spiral galaxies	Diameters of galaxies	100 Mpc
	Brightest galaxies in clusters	Largest galaxies in clusters	4000 Mpc
Clusters of galaxies	Luminosity function of cluster galaxies	Cluster core size	2000 Mpc

being used. The absolute magnitude M is defined as the magnitude the source would have at a distance of 10 pc, so

$$M = A - 2.5 \lg P + 87.45. \tag{3.3}$$

The absolute magnitudes of the Local Group of galaxies were shown in Table 1.1 (p. 4).

The *distance modulus* is then defined as $m - M$, and it can easily be seen from eqns (3.1)–(3.3) that it is related to luminosity distance by

$$m - M = 5 \lg d_{\text{lum}} - 87.45, \tag{3.4}$$

where d_{lum} is in metres.

The three primary distance indicators, which can be calibrated within our Galaxy and which are visible in nearby external galaxies, are: (i) RR Lyrae variable stars, with periods ranging from a few hours to a day; (ii) novae, eruptive stars with a characteristic pattern of rapidly brightening and slowly dimming light; (iii) Cepheid variable stars, with periods from 2 to 200 days and a well-determined relationship between period and luminosity. The range of these and other cosmological distance indicators is given in Table 3.1. In an important development for the distance scale, the Hubble Space Telescope has extended the range of the Cepheid method as far as the Virgo cluster.

Luminosity distances for different types of object give a ladder of distance reaching out to cosmological dimensions. However, it is a ladder that becomes shakier the further out it reaches. Only for galaxies in the Local Group (see Table 1.1, p. 4) can distances be measured by many independent methods.

One of the most promising distance indicators is supernovae, which can be seen in quite distant galaxies (up to 400 Mpc). By measuring the visible flux, the effective temperature, and the velocity of expansion of the light-emitting surface (from the Doppler spreading of spectral lines), a direct estimate of distance is obtained. This method has been applied to both Type Ia supernovae (white dwarfs in binary systems which blow up when a large mass of gas is dumped on them from their companion) and Type II supernovae (exploding massive stars).

Gravitationally lensed systems (see p. 70) can provide an ingenious estimate of distance if the lensed object varies its light output and the change is seen in two lensed images. Provided the lensing geometry is well understood, the time delay between the events in the two images provides a purely geometrical estimate of distance.

The diameter distance d_{diam} is defined by assuming that the apparent angle θ subtended by an object of intrinsic size l varies inversely with distance:

$$d_{\text{diam}} = l/\theta. \tag{3.5}$$

It can be obtained for only a few nearby stars, but can be used for more extended objects, ranging from planetary nebulae and globular clusters to clusters of galaxies. What is needed here is a class of object with a small dispersion in the intrinsic size l. Table 3.1 shows the contribution of diameter distance methods to the cosmological distance scale.

An interesting application of the diameter distance uses the Sunyaev–Zeldovich effect, in which the hot gas in clusters Compton-scatters photons from the microwave background radiation, causing a dark patch in the microwave background at frequencies below the peak and a bright patch at frequencies above it. By mapping the region at microwave and X-ray wavelengths, the properties of the gas cloud can be determined and the diameter distance of the cluster estimated.

Some empirical methods which have been very widely used to measure distances to galaxies, and have allowed the local velocity field to be mapped, are the *Tully–Fisher* method in spiral galaxies, based on a correlation between galaxy luminosity and maximum rotation speed, and the $D_n - \sigma$ method in elliptical galaxies, based on a correlation between galaxy diameter and velocity dispersion.

Some of the uncertainties which arise from these distance methods are well illustrated from our normal visual experience. The human eye uses parallax distance out to about 5 m, and diameter distances beyond that. However, on the road at night the only way of deciding the distance (and hence the speed) of a distant motor bike is by luminosity distance, but there is the possibility of confusion with a (much nearer) bicycle. For a car at night, diameter distance can be estimated from the angle subtended by the head or tail lights, but there is considerable error involved due to the great range in width between a mini and a juggernaut.

For objects travelling much slower than the speed of light in a Euclidean space all these definitions of distance give the same answer. In a curved space, discrepancies between the different methods may help to determine the curvature.

3.3 The redshift

If we look for some particular spectral lines in the Sun's spectrum, e.g. the sodium doublet at 5900 Å, we will find that the wavelengths of these and all the other spectral lines are in general shifted by a small amount. This is due to various factors:

(a) The rotation of the Earth about its axis results in a small Doppler shift of spectral lines, depending on the time of day. In the morning the observer has a component of motion towards the Sun, so the frequencies appear to be shifted to higher frequencies, towards the blue end of the spectrum. Similarly in the afternoon there is a shift towards the red. The amount is minute, at most 1.4 parts in a million.

(b) The Sun's own rotation, and circulatory motions in the Sun's surface layers, cause a shift of the same order of magnitude, the direction depending on which part of the Sun you are looking at.

(c) According to Einstein's general theory of relativity there will be a *gravitational redshift* by an amount GM/Rc^2, where M and R are the mass and radius of the Sun, and G is the gravitational constant. Essentially the photons have to do work in climbing out of the Sun's gravitational field, lose energy, and so end up at a lower frequency than when they set off. The fractional shift

$$\Delta\nu/\nu_e = (\nu_e - \nu_o)/\nu_e,$$

where ν_e and ν_o are the emitted and observed frequencies, again turns out to be about one part in a million.

(d) Now imagine that we look at another star in our own Galaxy. In addition to the shifts (a)–(c) above, there may be a far more significant shift due to the motion of the star round the Galaxy. The Sun, like the majority of stars in the plane of the Milky Way, is moving in a roughly circular orbit round the Galaxy, and has a circular velocity of 250 km s^{-1}. Some stars are weaving in and out of the galactic plane with speeds of the same order. The net result is that frequency shifts up to

$$|\Delta\nu/\nu_e| \sim 10^{-3} \tag{3.6}$$

may be expected.

Finally, what happens when we look at the integrated spectrum of another galaxy? First, since the light from a galaxy is made up of light from many stars, we will expect the spectral lines to be spread out by about the amount given by eqn (3.6). Secondly, since the galaxies have some relative motion with respect to each other (nearby ones affect each other gravitationally, for example), we would expect to find some moving towards us, some moving away, i.e. some blueshifted, some redshifted.

We do indeed find this for the nearest galaxies, e.g. the members of the Local Group of galaxies mentioned in Section 1.1. Typical (so-called 'peculiar') velocities are a few hundred kilometres per second. But when we start to look at more distant galaxies, as determined by the methods of Section 3.2, a remarkable fact emerges (discovered by Hubble and Lundmark in the 1920s and strongly confirmed by modern work). Almost all the frequency shifts are redshifts and *the redshift increases linearly with distance*. The slope of this line determines a constant with the dimensions of distance, which we shall call the *Hubble distance* and write as $c\tau_0$, where τ_0 is the *Hubble time* (the time for light to travel a Hubble distance). If we write z for the redshift.

$$z = \Delta\nu/\nu_o = (\nu_e - \nu_o)/\nu_o,$$

then

$$z \cong \frac{d}{c\tau_0} + \frac{\upsilon_{pec}}{c} \tag{3.7}$$

where d is the distance of the galaxy, υ_{pec} is its peculiar velocity in the line of sight, and $d/c\tau_0$ is the *cosmological redshift*.

Except for nearby galaxies, the peculiar velocity can be neglected, so strong is the effect of the cosmological redshift. If the redshift is interpreted as a Doppler shift, then the galaxies are receding from us in every direction, with a velocity that increases with distance from us (Figs 3.1 and 3.2). In fact in this case

$$\begin{aligned} \upsilon &\cong zc \cong d/\tau_0, \qquad \text{provided } \upsilon \ll c, \\ &= H_0 d, \end{aligned} \tag{3.8}$$

where $H_0 = \tau_0^{-1}$ is called the *Hubble constant* . If τ_0 is given in years, H_0 is in $(\text{years})^{-1}$. However, you will often find it given in units of km s^{-1} Mpc^{-1}. Current estimates lie in the range H_0 = 50–100 km s^{-1} Mpc^{-1}, corresponding to $\tau_0 = 1\text{–}2 \times 10^{10}$ years and

$$c\tau_0 = 1\text{–}2 \times 10^{26}\ \text{m} = 3000\text{–}6000\ \text{Mpc}.$$

Clearly the time for a galaxy at distance d travelling at velocity υ to travel a further distance d is, by eqn (3.8), $d/\upsilon = \tau_0$, independent of d. Thus the Hubble time is a measure of the expansion time of the universe, the time for the universe to double its size expanding at the present rate.

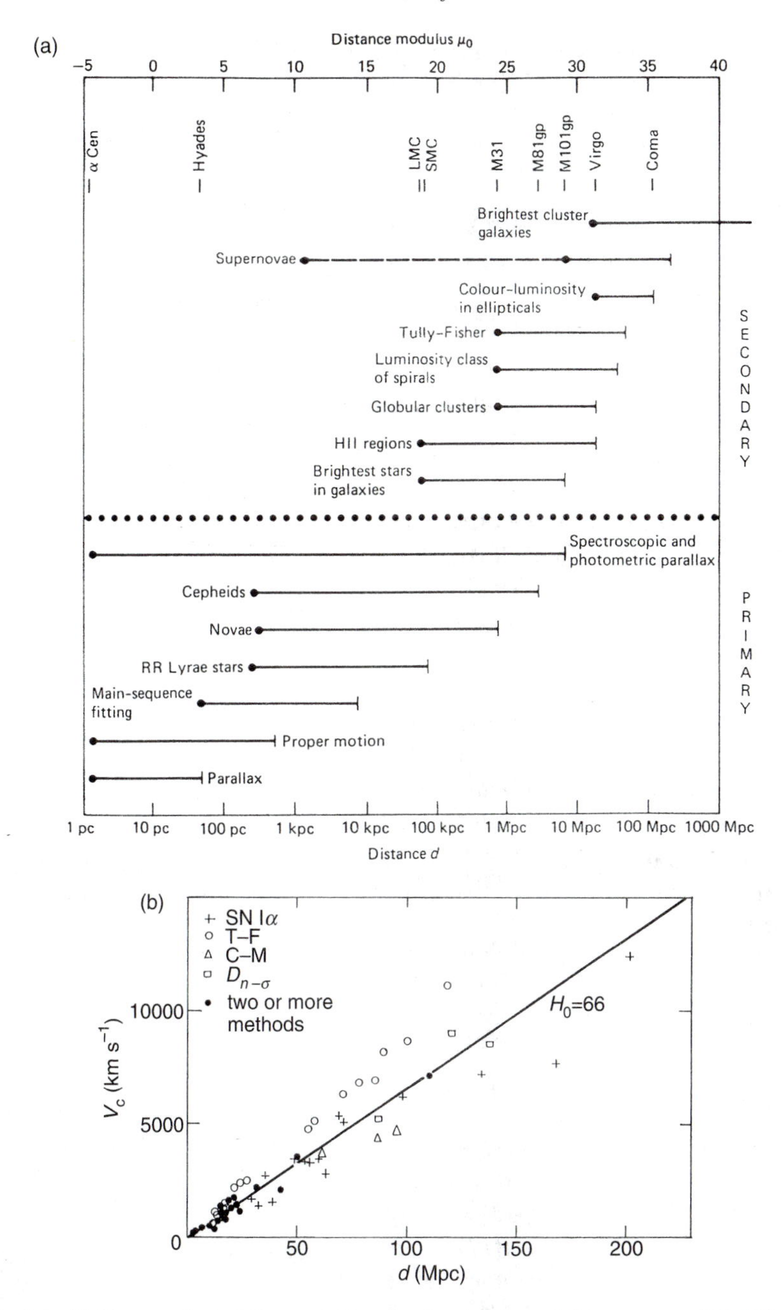

Fig. 3.1 (a) The cosmological distance ladder, showing the range of distances over which different distance methods have been applied. From *The cosmological distance ladder* (Rowan-Robinson 1985). (b) The velocity–distance relation, with different symbols for different methods. The Hubble constant probably lies in the range 50–80 km s^{-1} Mpc^{-1}, with a current best average value in the middle of this range. From Rowan-Robinson (1988) *Space Science Reviews* **48**, p. 1.

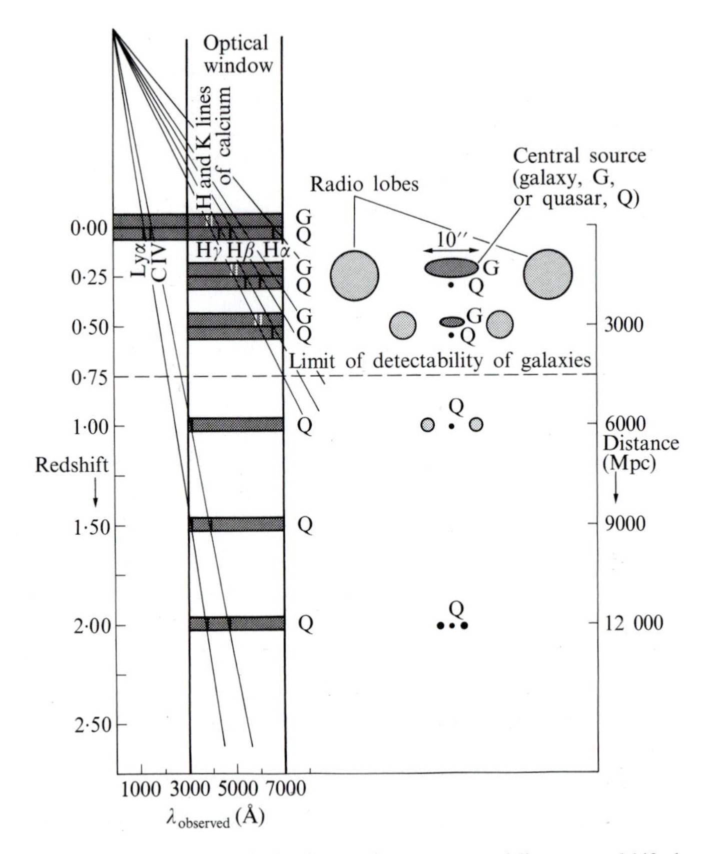

Fig. 3.2 Illustration of the way typical galaxy and quasar spectral lines are redshifted across the visible wavelength band, and of the way this redshift increases with distance, indicated by the angular size and separation of the radio components and by the angular size of the galaxy (the quasar appears point-like at all distances).

3.4 Isotropy

Galaxies

We saw in Chapter 1 that the distribution of the nearest galaxies on the sky is far from isotropic, due to our membership of the Local Group of galaxies and proximity to the Virgo cluster. But on the large scale the distribution of galaxies seems to be fairly isotropic, once allowance has been made for the obscuration due to dust in our Galaxy. The exact nature of these tiny (0.1 μm diameter) dust grains responsible for absorbing and scattering light within the Galaxy is uncertain, but they are probably composed of silicates, graphite, or silicon carbide, depending where they originate from. In denser gas clouds, mantles of H_2O ice and, possibly, complex hydrocarbon molecules form on the grains.

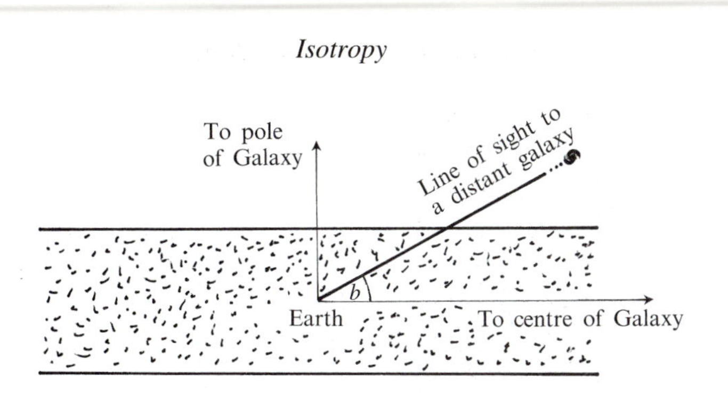

Fig. 3.3 Part of the Galactic disc near the Earth. The line of sight to a distant galaxy has to traverse a distance h cosec b of the dusty gas in the disc, where $2h$ is the thickness of the disc and b is the Galactic latitude.

The grains appear to be distributed throughout the gas pervading the thin disc of our Galactic plane. The result is that a line of sight close to the plane passes through a much longer column of dust than a direction normal to the plane. For a uniform distribution of dust throughout a slab of thickness $2h$, the path length that light from outside the slab has to traverse is h cosec b, where b is the Galactic latitude of the source (Fig. 3.3). In fact the dust concentration varies with distance from the Galactic plane in a roughly inverse exponental way. However, in the approximation that the concentration depends only on distance from the plane over the line of sight, the cosec law still applies.

The flux from a distant source is then reduced from S_0 to

$$S = S_0 \exp(-kh \operatorname{cosec} b), \tag{3.9}$$

where k denotes the extinction per unit length (i.e. $\mathrm{d}S/\mathrm{d}x = -kS$). In terms of magnitudes,

$$m = m_0 + 1.086\, kh \operatorname{cosec} b. \tag{3.10}$$

Thus if we have a telescope capable of seeing galaxies down to a particular value of m, the corresponding limiting value of m_0 will depend on b. For smaller values of b we cannot see such distant objects as we can towards the Galactic pole ($b = 90°$). This in turns means that we will not see so many galaxies per square degree, on average.

The actual distribution of galaxies, in Galactic coordinates (analogous to latitude and longitude, with the Galactic plane as equator), is shown in Fig. 3.4. This is derived from Hubble's counts of the numbers of galaxies per square degree in different directions. Within the 'zone of avoidance' close to the Galactic plane, no galaxies are seen (see also Fig. 3.5).

Naturally it is hard to say much about the isotropy of the universe from these data, but once a correction is applied for obscuration the distribution is certainly *compatible* with isotropy, apart from the tendency already mentioned for galaxies to occur in clusters. Assuming the distribution is highly isotropic (not a very good assumption), counts of galaxies can be used to map the Galactic dust obscuration.

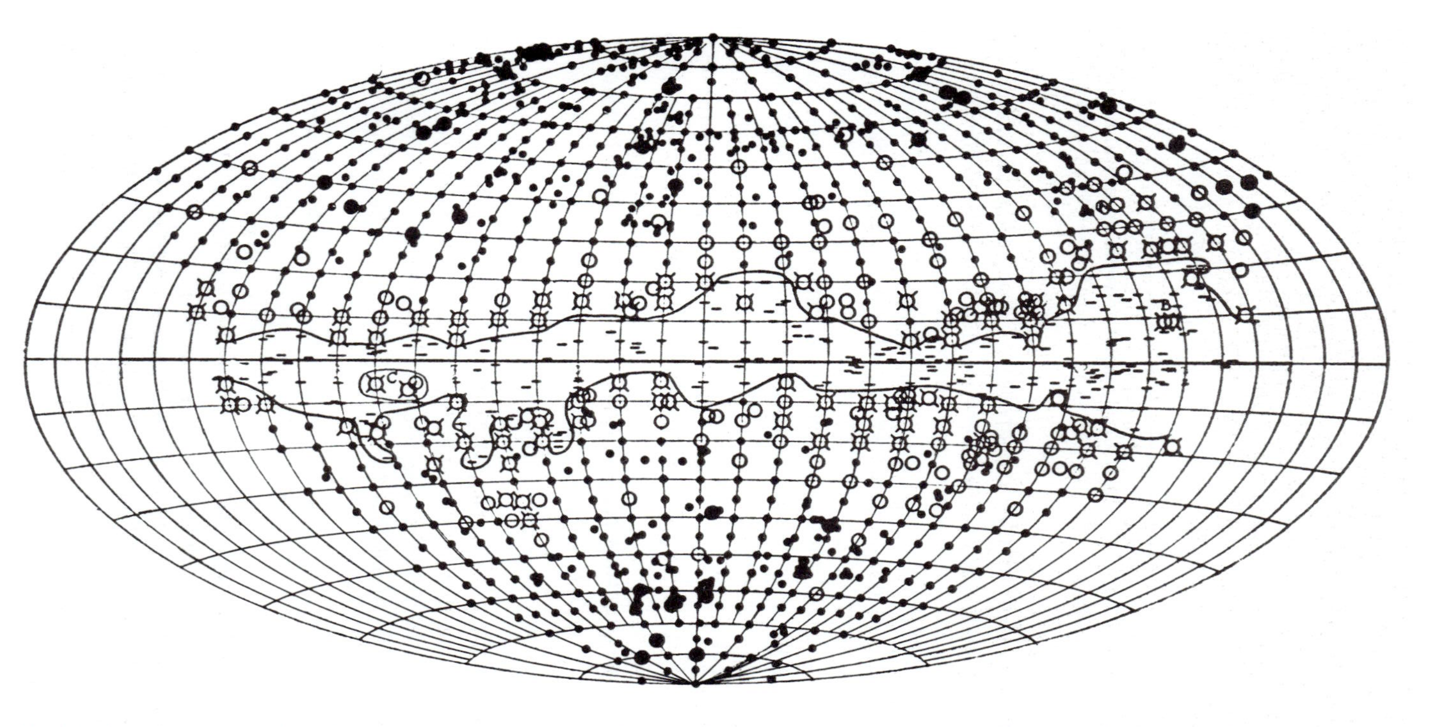

Fig. 3.4 The distribution of Hubble's galaxy counts to 20th magnitude on the sky in galactic coordinates. Large filled circles denote fields where an above-average number of galaxies were counted, large open circles where the number was below average, and a dash where none at all could be seen. Near the Galactic plane there is a 'zone of avoidance' where almost no galaxies are seen, owing to obscuration by dust.

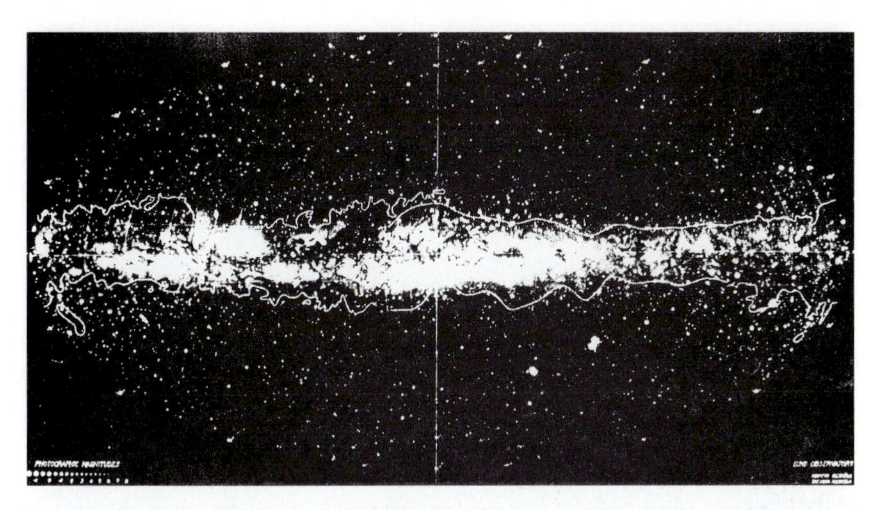

Fig. 3.5 Composite photograph of the Milky Way by the Lund Observatory, with a contour of constant number of galaxies per square degree from the Lick and Harvard counts superimposed. The contour corresponds to a low number per square degree to illustrate how the zone of avoidance follows the regions of high star (and hence gas and dust) density in the Galaxy.

The survey made by the IRAS satellite at far-infrared wavelengths has allowed both the accurate mapping of the Galactic dust distribution and the mapping of the galaxy distribution within 500 million light years free from the effects of dust obscuration. Within this volume, galaxies are concentrated into a number of large clusters, but are otherwise broadly isotropically distributed.

Clusters of galaxies

Abell and others have made catalogues of 'rich' clusters of galaxies, i.e. those with a specified minimum number of bright members within a specified volume. Their distribution on the sky shows evidence for large-scale 'superclustering' but is again compatible with isotropy on the very largest scales, once the correction for obscuration has been applied.

Radio sources

The vast majority of the bright radio sources in directions away from the Galactic plane are extragalactic, either galaxies or quasars. There is no Galactic extinction of radio waves, except at very low frequencies, so we can test directly for isotropy. So far, down to the faintest sources at present detectable, the distribution of radio sources on the sky is isotropic to a few per cent. Such discrepancies as have arisen between one survey and another have usually turned out to be errors in the flux-density scale.

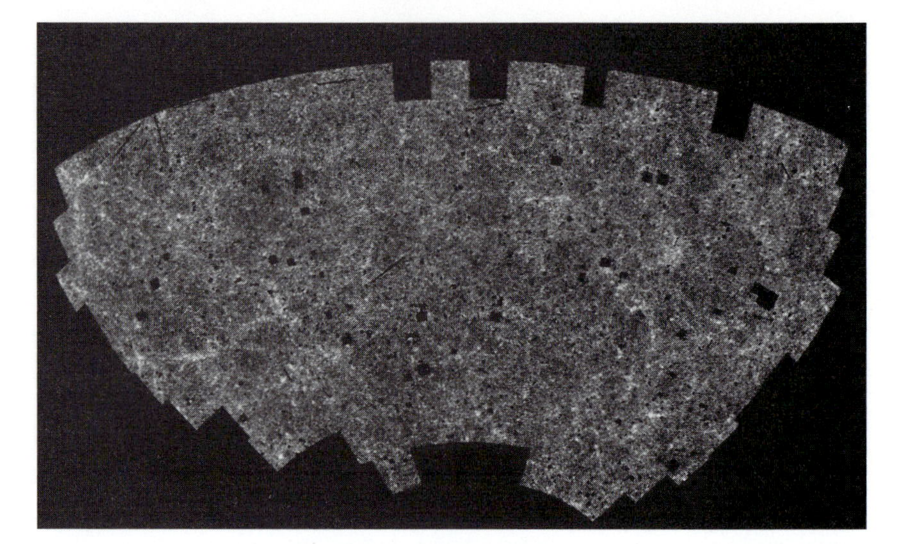

(a)

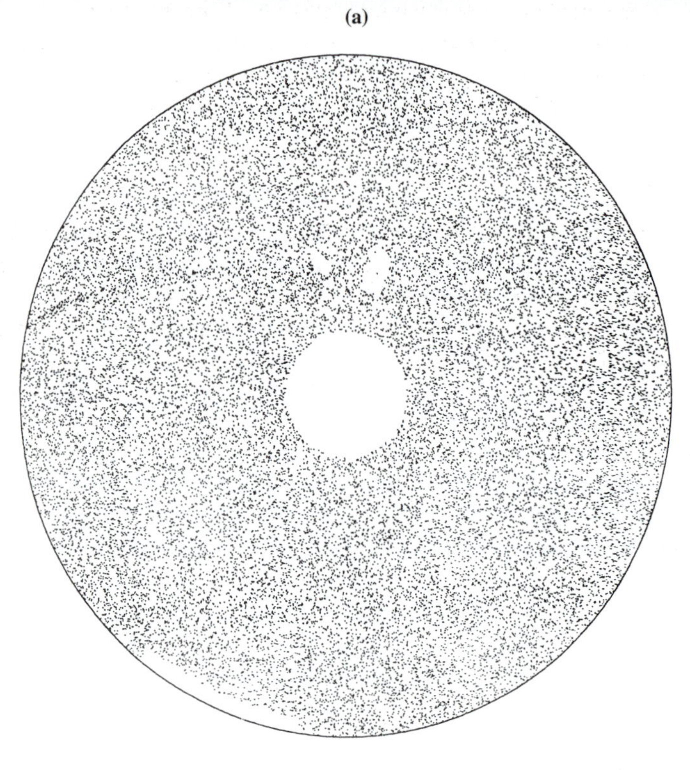

(b)

Fig. 3.6 (a) Distribution of galaxies at the south Galactic pole from a survey made from UK Schmidt telescope plates using the APM automatic plate measuring machine. Photograph courtesy of G. Efstathiou. (b) Distribution on sky of the 31 000 brightest 6 cm radio sources in the northern sky. From Gregory and Condon (1991) *Astrophysical Journal Supp.*, **75**, p. 1011.

Because radio surveys reach to very great distances, the isotropy of radio sources is the strongest evidence we have of the isotropy of the galaxy distribution on large scales.

The microwave background radiation

So far we have looked at the distribution on the sky of matter and although no clear anisotropy has emerged, the degree of isotropy we can claim is no better than a few per cent. Far more significant evidence for the isotropy of the universe on the large scale comes from the microwave background radiation (Section 1.8). At centimetre wavelengths, the COBE satellite has shown that this radiation is isotropic on the large scale to 0.001 per cent (1 part in 10^5), apart from the effect of the Earth's motion (see Section 5.5).

This is a superb confirmation of Einstein's 1917 postulate of an isotropic universe and provides us with a strong basis for the isotropic models on which most cosmology today is based. This isotropy dates from long times in the past, and places severe limits on any anisotropic models of the universe.

The X-ray background in the 2–18 keV band in directions away from the Galactic plane has been shown to be isotropic to 1 per cent. The origin of the X-ray background is still a matter of some controversy (see Section 7.10).

3.5 Uniformity

Newton realized that if the stars were distributed with uniform number density throughout the universe, as had been proposed by Digges and Bruno, then this could be tested by counting the number of stars as a function of their observed flux.

For if a set of sources, all of the same luminosity P, are distributed uniformly with number density η, then the number per steradian out to distance r is

$$N(r) = \eta r^3/3,$$

while their observed flux is

$$S = P/r^2.$$

The number of sources per steradian brighter than S is therefore

$$N(S) = (\eta P^{3/2}/3)S^{-3/2} \tag{3.11}$$

or

$$\begin{aligned} \lg N &= A - 1.5 \lg S \\ &= B + 0.6m \end{aligned} \tag{3.12}$$

by eqn (3.2), where A and B are constants.

Newton's attempts to apply this test of uniformity failed because he had no adequate way of estimating the fluxes of stars. Later, William Herschel used this test to show that the stars of our Galaxy are in fact distributed in a disc.

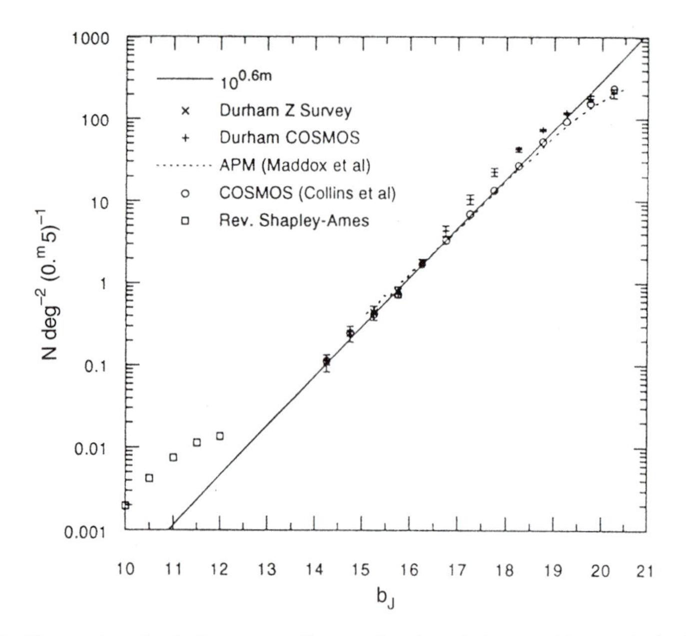

Fig. 3.7 The number of galaxies per steradian as a function of photographic magnitude showing, at least for the relatively bright magnitudes included here, good agreement with the expected $d(\lg N)/dm + 0.6$ relation. Figure by T. Shanks.

For distant galaxies we do not expect eqn (3.12) to hold exactly, since the redshift, whatever its interpretation, will affect the observed fluxes. But for brighter galaxies, eqn (3.12) does indeed hold approximately (Fig. 3.7). For radio sources, on the other hand, a significantly steeper slope is found (Fig. 3.8), suggesting that the number density or luminosity of these sources was greater in the past. However, redshift effects cannot be neglected in interpreting radio-source counts.

3.6 Olbers' paradox

Halley, and later Cheseaux and Olbers, realized that important cosmological information is contained in the fact that the sky is dark at night.

Suppose a population of sources of luminosity P has a number density η Consider a spherical shell of radius r, thickness dr, centred on the Earth. The number of sources in one steradian of the shell is

$$dN = \eta r^2 dr,$$

and their flux is

$$S = P/r^2$$

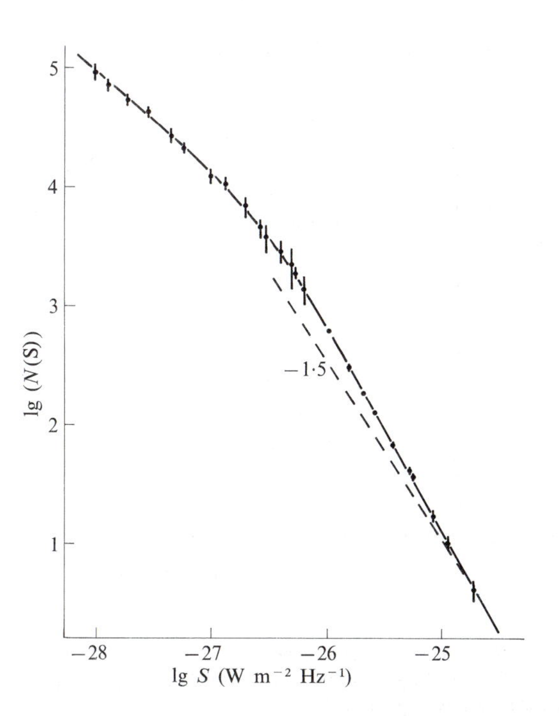

Fig. 3.8 Radio-source counts, lg N against lg S, at a frequency of 408 MHz. Even for the brightest sources, the curve rises more steeply than the -1.5 predicted (eqn. (3.12)) if all redshift and evolutionary effects are neglected.

The intensity of the integrated light from these sources is

$$dI = SdN = \eta P dr$$

and if we add the contribution from shells with $0 \leq r \leq R$, the total intensity is

$$I = \eta PR \tag{3.13}$$

Clearly if we let $R \to \infty$, then $I \to \infty$. (Actually the stars would eventually start to block out more distant light, so the intensity would tend to the average surface brightness of the stars—something comparable to the surface brightness of the solar disc.) Since the intensity of the integrated light from galaxies has to be much less than the intensity of the Milky Way, we can deduce that we receive light only from galaxies with a distance R less than some maximum distance R_{max}. The simplest reason for this would be that the universe is of finite age. An infinitely old expanding universe would be another way of resolving this paradox.

This 'paradox' can also be stated in a thermodynamic form: why is the universe so cold? Again, in a universe of finite age, we see that there may not have been time for the stars to heat the rest of the matter up.

3.7 Evidence for a universe of finite age

Several pieces of evidence suggest that we may live in a universe of finite age.

1. The proportions of different isotopes of radioactive elements allow quite accurate dating of different materials. For example, the oldest rocks on the Earth, the Moon, and in meteorites have ages of about $(4.5 \pm 0.1) \times 10^9$ years, in good agreement with the age of the Sun estimated from calculations of its structure and evolution, 5×10^9 years. Applied to the material in our Galaxy as a whole, radioactive dating gives an age of 1.1–1.8×10^{10} years, again agreeing well with estimates of the ages of the oldest stars, from stellar evolution theory 1.2–1.6×10^{10} years.

2. The ages of other nearby galaxies are of the same order, independent of galaxy type.

3. If the cosmological redshift is due to expansion of the universe, then the age of the universe will be of order the Hubble time, 1–2×10^{10} years. The similarity in the ages of galaxies in our neighbourhood and the expansion time-scale of the universe strongly suggests a universe of finite age in which galaxies formed early on. However, it is hard to eliminate the possibility that the part of the universe we have studied so far is merely some local fluctuation, or that the universe has had a cyclical history.

3.8 Evidence for a 'fireball' phase

The existence of an early phase of the universe's history in which radiation played a dominant role (the 'fireball' phase) is the simplest explanation of the 2.7 K blackbody background radiation.

A second piece of evidence for such a phase is that the stars in our Galaxy appear to have been formed with an initial composition of 76 per cent hydrogen, 24 per cent helium, by mass. Almost all the remaining elements can be formed within the interiors of stars, so it is natural to suppose that the universe started as pure hydrogen and that the helium was formed in some pre-stellar phase, i.e. the fireball. While it is possible to construct a theory of the early stages of our Galaxy's history in which very massive stars formed, evolved rapidly, and exploded, dispersing helium, the most natural explanation is that the helium was formed through nuclear reactions in the early fireball phase of the universe.

3.9 Problems

3.1 Use eqn (3.10) to work out how many magnitudes an object in the direction (a) of the Galactic pole, (b) Galactic latitude $b = 30°$, (iii) $b = 10°$, is dimmed, given that there are about 20 magnitudes of extinction towards the Galactic centre (8 kpc distant) and that $h = 150$ pc.

3.2 A spectrograph operates in the range 5000–9000 Å. Determination of the redshift of emission-line galaxies or quasars depends on the detection of the

Lyman α (1216 Å), Carbon IV (1549 Å), Balmer Hα (4861 Å) or Balmer Hβ (6563 Å) lines. What ranges of redshift will be covered by (i) one, (ii) two detectable spectral lines?

4
The big-bang models

4.1 The substratum and fundamental observers

To describe the properties of a gas we do not need to study the behaviour of individual atoms and molecules. Instead we define various macroscopic quantities—density, pressure, temperature—and study the relations between these.

In the same way, we make no attempt to incorporate individual galaxies, or clusters of galaxies, into our description of the universe as a whole. We imagine the matter in the universe smeared out into an idealized, smooth fluid, which we call the *substratum*.

We call an observer who is at rest with respect to this substratum *a fundamental observer*. If the substratum is in motion, then we say that fundamental observers are *co-moving* with it. We are interested in obtaining the picture that fundamental observers have of the universe as a function of time. We call such a set of pictures a *cosmological model*.

In practice we will often identify fundamental observers with galaxies, and will treat an observer at the centre of our Galaxy as one, although in fact individual galaxies will have some 'peculiar' velocity with respect to the substratum. (The peculiar velocity of our Galaxy has been measured through observations of the microwave background radiation.) But for the moment we shall assume that when we have corrected for the rotation of the Earth, its orbit round the Sun, and the motion of the whole solar system round the Galaxy, then we are receiving a fundamental observer's view of the universe.

4.2 The cosmological principle

It is evident that in the post-Copernican era of human history, no well-informed and rational person can imagine that the Earth occupies a unique position in the universe. We shall call this profound philosophical discovery the *Copernican principle*, although the first clear statement of it is due to Giordano Bruno. The discovery of millions of stars like the Sun, of other possible planetary systems, and of galaxies similar to our own Galaxy, all help to convince us of the truth of the Copernican principle. Bruno himself knew of no such evidence, so his affirmation was more of a poetic, psychological, and even political truth.

We now explore the consequences of a much more powerful assumption, the *cosmological principle*: the universe as seen by fundamental observers is homogeneous and isotropic.

By *homogeneity*, we mean that every fundamental observer sees the same general picture of the universe as a function of time. Every fundamental observer is equivalent to every other and, in particular, the universe as seen by any fundamental observer looks the same as the universe as seen from Earth. The hypothesis of homogeneity can never be strictly tested, for even if advanced civilizations in distant galaxies transmitted their cosmological knowledge to us, it would always be out of date by the time it arrived. The power of the hypothesis is that our own observations are then all we need to test a cosmological model.

By *isotropy*, we mean that the universe looks the same to a fundamental observer in whichever direction in the sky he or she looks. Observations in one direction only are then sufficient to test a cosmological model. The hypothesis is only roughly (~ 3 per cent) tested for the matter in the universe, but it is very accurately (0.001 per cent) satisfied for the microwave background radiation (Section 3.4).

The cosmological principle satisfies the Copernican principle in almost the strongest possible way. Considerable theoretical effort has gone into investigating models which are homogeneous but anisotropic. It can be shown that isotropy, together with the Copernican principle, implies homogeneity. The actual universe is clearly highly inhomogeneous on the small scale and only rough limits can be put on its homogeneity on the large scale. The mathematical difficulties of describing and discovering the properties of an inhomogeneous universe are formidable. We therefore have to hope that the homogeneous models described in this chapter are a reasonable approximation on the large scale, bearing in mind that there is so far no successful explanation of why this idealized state of affairs should hold (see Section 8.6).

An immediate consequence of homogeneity is the existence of a universal *cosmical time*, which we shall denote by t. For since all observers see the same sequence of events in the universe, they can synchronize their clocks by means of these events.

4.3 Newtonian cosmology

Newtonian dynamics and gravitation can be used to construct models of the universe satisfying the cosmological principle. However, Newtonian cosmology is not strictly self-consistent and the justification of the models depends on a result from the general theory of relativity, Birkhoff's theorem, as we shall see below.

In Newtonian cosmology, the cosmical time t can be identified with the uniform ever-flowing universal Newtonian time.

Consider a fundamental observer O. He or she sets up coordinates with themselves at the origin, and observes the physical properties of the matter at a general point P at time t, the position of P being given by $\overrightarrow{\mathrm{OP}} = \boldsymbol{r}$. Naturally O finds that the velocity, density, and pressure at P are functions of position and time, $\boldsymbol{v}(\boldsymbol{r}, t)$, $\rho(\boldsymbol{r}, t)$, $p(\boldsymbol{r}, t)$.

Now consider a second fundamental observer O′, distant $\overrightarrow{\mathrm{O'P}} = \boldsymbol{r}'$ from P. He or she finds $\boldsymbol{v}'(\boldsymbol{r}', t)$, $\rho(\boldsymbol{r}', t)$, $p(\boldsymbol{r}', t)$. Note that in Newtonian theory, the density

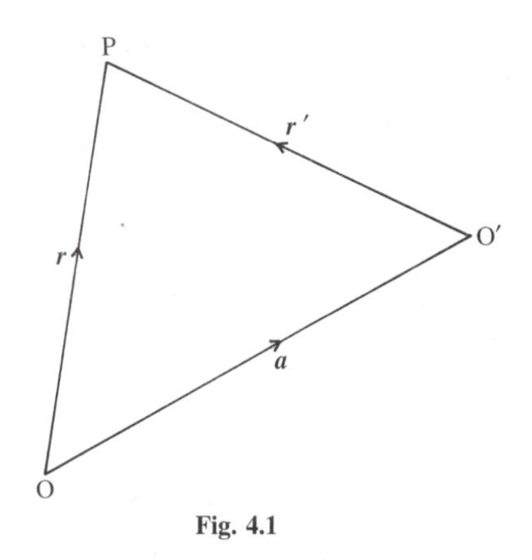

Fig. 4.1

and pressure at any point are the same for all observers (we say they are *invariants* with respect to a change of coordinate system). To satisfy the cosmological principle we now demand that $\boldsymbol{v}'$, ρ, and p should be the same function of $(\boldsymbol{r}', t)$ as $\boldsymbol{v}$, ρ, and p are of $(\boldsymbol{r}, t)$, otherwise O′ and O would have different pictures of the universe.

At a fixed time t, let $\overrightarrow{OO'} = \boldsymbol{a}$, so that the velocity of O′ as seen by O is $\boldsymbol{v}(\boldsymbol{a})$ (we shall ignore the dependence on t for the moment). From Fig. 4.1 we can see that

$$\boldsymbol{r}' = \boldsymbol{r} - \boldsymbol{a}, \tag{4.1}$$

by the vector law of addition applied to the triangle OO′P (Fig. 4.1). It also follows that

$$\boldsymbol{v}'(\boldsymbol{r}') = \boldsymbol{v}(\boldsymbol{r}) - \boldsymbol{v}(\boldsymbol{a}), \tag{4.2}$$

by the vector combination of velocities.

Now eqn (4.1) implies that

$$\boldsymbol{v}'(\boldsymbol{r}') = \boldsymbol{v}'(\boldsymbol{r} - \boldsymbol{a}), \tag{4.3}$$

and homogeneity requires that

$$\boldsymbol{v}'(\boldsymbol{r} - \boldsymbol{a}) = \boldsymbol{v}(\boldsymbol{r} - \boldsymbol{a}), \tag{4.4}$$

since O′ and O have to see exactly the same picture of events. Collecting eqns (4.2)–(4.4) together:

$$\boldsymbol{v}(\boldsymbol{r} - \boldsymbol{a}) = \boldsymbol{v}(\boldsymbol{r}) - \boldsymbol{v}(\boldsymbol{a}). \tag{4.5}$$

Homogeneity also tells us that the way density and pressure vary with position as measured by O must be the same as the way they vary with position as seen by O′, i.e.

$$\rho(\boldsymbol{r}') = \rho(\boldsymbol{r}); \quad p(\boldsymbol{r}') = p(\boldsymbol{r}),$$

and then eqn (4.1) implies that

$$\rho(\boldsymbol{r} - \boldsymbol{a}) = \rho(\boldsymbol{r}); \quad p(\boldsymbol{r} - \boldsymbol{a}) = p(\boldsymbol{r}). \tag{4.6}$$

Since O, O′, P are arbitrary so also are $\boldsymbol{r}$, $\boldsymbol{r}'$, $\boldsymbol{a}$, and therefore eqns (4.6) imply that p and ρ must be independent of position.

The general solution of the three equations (4.5) can be shown to be

$$v_i(\boldsymbol{r}, t) = \sum_{k=1}^{3} a_{ik}(t) x_k, \quad \text{for } i = 1, 2, 3, \tag{4.7}$$

where v_i denotes the ith component of $\boldsymbol{v}$; x_k denotes the kth component of $\boldsymbol{r}$; and $a_{ik}(t)$, $i, k = 1, 2, 3$, are nine arbitrary functions of the time t. (Writing out the first of the three equations (4.7) in full:

$$v_1(x_1, x_2, x_3, t) = a_{11}(t)x_1 + a_{12}(t)x_2 + a_{13}(t)x_3).$$

For the flow field represented by eqn (4.7) to be isotropic we must have

$$a_{ik} = 0, \quad i \neq k,$$

and

$$\begin{aligned} a_{11} &= a_{22} = a_{33} \\ &= H(t), \end{aligned}$$

say. Therefore

$$v_1 = H(t)x_1, \quad v_2 = H(t)x_2, \quad v_3 = H(t)x_3$$

or

$$\boldsymbol{v} = H(t)\boldsymbol{r}. \tag{4.8}$$

This equation tells us that the velocity of any particle moving with the substratum is either zero or is directed radially away from or toward us with a velocity proportional to distance. In other words we have a natural explanation of the Hubble law, with the cosmological redshift being interpreted as a Doppler shift. And since this velocity field satisfies the cosmological principle, any other fundamental observer sees exactly the same picture: every particle moving with the substratum has a purely radial velocity, proportional to its distance from the observer.

Eqn (4.8) can be integrated, since $\boldsymbol{v} = \mathrm{d}\boldsymbol{r}/\mathrm{d}t$, by writing

$$H(t) = \frac{1}{R(t)} \frac{\mathrm{d}R(t)}{\mathrm{d}t}$$

giving

$$\frac{\mathrm{d}\boldsymbol{r}}{\mathrm{d}t} = \frac{1}{R(t)} \frac{\mathrm{d}R}{\mathrm{d}t} \boldsymbol{r},$$

which has the solution

$$\begin{aligned} \boldsymbol{r} &= R(t) \times \text{ a constant vector} \\ &= \frac{R(t)}{R_0} \boldsymbol{r}_0 \end{aligned} \tag{4.9}$$

where $R_0 = R(t_0)$, so that $\boldsymbol{r} = \boldsymbol{r}_0$ at epoch $t = t_0$. $R(t)$ is called the *scale factor* of

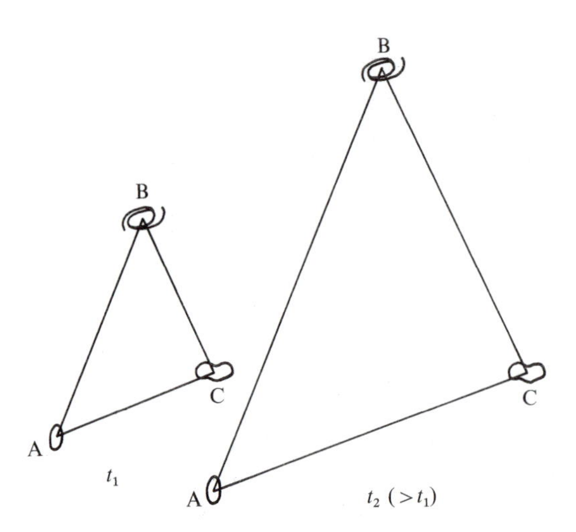

Fig. 4.2 Illustrating the way all distances are scaled up by the same factor as time changes.

the universe, since as time proceeds all distances are simply scaled up by this factor (Fig. 4.2). The only motions permitted by the cosmological principle are a simple isotropic expansion or contraction of the whole universe.

Since the volume will be proportional to $R^3(t)$, the density of matter

$$\rho(t) \propto R(t)^{-3}, \tag{4.10}$$

i.e.

$$\rho(t) = \rho_0 R_0^3 / R^3(t), \tag{4.11}$$

where $\rho_0 = \rho(t_0)$. We shall generally take the reference epoch $t = t_0$ to be the present epoch, and when quantities have a subscript zero it will denote their values at the present epoch. Since at the present epoch the energy density of radiation in the universe has a negligible dynamical effect, we shall for the moment take the pressure p to be zero. This assumption will break down in the early stages of the big bang (see Chapter 5).

To find the form of the function $R(t)$, consider a spherical shell of particles co-moving with the substratum, centred on the observer O. From eqns (4.8) and (4.9), particles A, B, C, D on this spherical shell are receding with velocity $\dot{R}\mathbf{r}_0/R_0$, and have acceleration $\ddot{R}\mathbf{r}_0/R_0$, where $\dot{R}$ and $\ddot{R}$ denote $(\mathrm{d}R/\mathrm{d}t)(t)$ and $(\mathrm{d}^2R/\mathrm{d}t^2)(t)$.

It is known in the Newtonian theory of gravitation that the gravitional force inside a uniform spherical shell of matter is zero. Thus if we could imagine the whole universe as a large sphere centred on O and divided up into thin concentric spherical shells, the shells exterior to the shell ABCD defined above will have no gravitational influence on the particles A, B, C, D. However we cannot

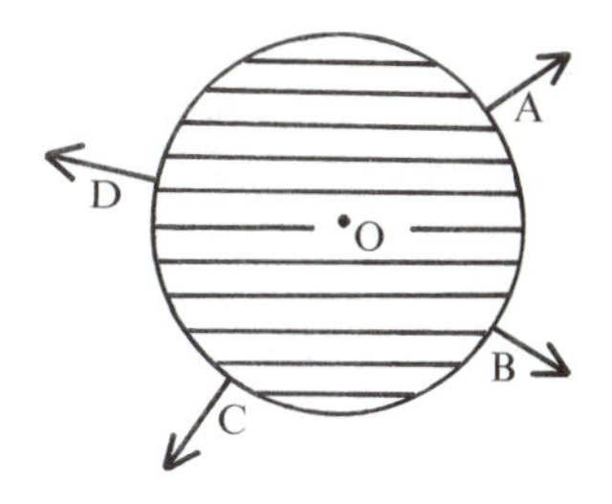

Fig. 4.3 According to O, the only force acting on A, B, C, D is the gravitational attraction of the shaded sphere, centred on O.

extrapolate this to an infinite universe, and to justify a Newtonian treatment of the cosmological problem we have to invoke a result from the general theory of relativity, Birkhoff's theorem. This states that the gravitational effect of a uniform medium external to a spherical cavity is zero. Then, according to O, the only force acting on the particles is the gravitational attraction of the matter interior to the shell ABCD (see Fig. 4.3). Thus by Newton's second law of motion, the force on a particle of mass m on this shell is

$$m\ddot{\boldsymbol{r}} = -\frac{4\pi G m \rho}{3}\boldsymbol{r}.$$

Substituting from eqn (4.9) and cancelling m,

$$\ddot{R}\frac{\boldsymbol{r}_0}{R_0} = -\frac{4\pi G \rho R \boldsymbol{r}_0}{3R_0}$$

so

$$\ddot{R} = -4\pi G \rho R/3. \tag{4.12}$$

This can be integrated, using eqn (4.11), to give

$$\dot{R}^2 = \frac{8\pi G \rho_0 R_0^3}{3R} - kc^2, \tag{4.13}$$

where k is an arbitrary constant, dimensionless if R is taken to have dimensions of distance. Before we look for solutions of this, we shall consider what kinds of model are consistent with the cosmological principle in Einstein's general theory of relativity.

4.4 The special and general theories of relativity

To explain these theories properly would need a whole book in itself, so here I shall give only a thumbnail sketch.

Newton postulated the existence of *inertial* frames of reference, in which the motion of a free particle (i.e. with no forces acting on it) would be a straight line.

The surface of the Earth is a good approximation to a Newtonian inertial frame, apart from the small effects of the Earth's rotation. Of course all particles are acted on by gravity, which in Newtonian theory is a force. Once you have found one inertial frame, all the others can be found by performing one or both of the operations (a) changing the origin, and changing the orientation of the coordinate axes; (b) moving with a uniform velocity with respect to the original frame.

Newtonian theory works very well for most purposes, but it breaks down when we consider the velocity of light from moving sources. It predicts that if we measure the velocity of light from a source moving towards us, the result should be the sum of the velocity of the source and the velocity of light itself. However, the Michelson–Morley experiment showed that in fact the velocity of light is independent of the velocity of the source. *Special relativity* takes this fact as its starting point, and makes a modification, which under normal circumstances is small, to the way physical quantities change under the operation (b) above. The modification becomes large when relative velocities close to the velocity of light are involved. A moving body appears to have its length in the direction of motion contracted, its time slowed down, and its mass increased, all by a factor

$$\gamma = (1 - \upsilon^2/c^2)^{-1/2},$$

where υ is its velocity. This factor becomes infinite as υ approaches c, so a body can never be accelerated past the speed of light. These effects have been accurately verified, in large particle accelerators, for example, as also has the famous prediction of the equivalence of mass and energy, summarized by $E = mc^2$.

So far we have dealt only with uniform relative velocities. We can deal with an observer who is accelerated with respect to an inertial frame by assuming that he or she sees the same picture of the world as an inertial observer instantaneously at rest with respect to him (the *acceleration* principle), but we no longer know what to do about gravity, since Newtonian theory is not compatible with special relativity. Einstein solved this problem with his *general theory of relativity*. This is based on the principle of equivalence, which stated most simply says that gravity disappears if you let yourself fall freely. Everything experiences the same acceleration under gravity (Galileo's experiment) so if you are falling too, the effects of gravity seem to vanish. At every point we can find a frame of reference in which special relativity holds locally (the local freely falling frame). Gravity is reduced to the status of a non-inertial force like centrifugal or coriolis force, which appears only because we have not chosen the right frame of reference to make our observations.

In relativity theory the idea of an event, something happening in a particular place at a particular time, plays a large role. An event can be characterized by the four coordinates (x_1, x_2, x_3, t). where the first three are spatial coordinates and the fourth represents time. We can think of events as points in the four-dimensional *space-time* continuum. If a light signal is emitted at event (x_1, x_2, x_3, t) and received at a nearby event $(x_1 + \mathrm{d}x_1, x_2 + \mathrm{d}x_2, x_3 + \mathrm{d}x_3, t + \mathrm{d}t)$ (Fig. 4.4) then we know that, according to the special theory of relativity.

$$\begin{aligned} ds^2 &= dt^2 - \frac{1}{c^2}(dx_1^2 + dx_2^2 + dx_3^2) \\ &= 0. \end{aligned} \tag{4.14}$$

If the two events are not linked by a light signal, then $ds^2 \neq 0$. The quantity ds is called the *interval* between the two neighbouring events, and in the special theory all inertial observers get the same answer for ds when they observe the same pair of events (this is obviously true when the two events are connected by a light signal, since the velocity of light is c in all inertial frames. so they all get the answer $ds = 0$). We say ds is an *invariant.*

ds is essentially the time interval measured by an observer who is present at both events. If $ds^2 > 0$ then an observer can, by choosing the right speed, be present at both events. In this frame of reference $dx_1 = dx_2 = dx_3 = 0$, so $ds = dt$. ds just measures the interval of time on a clock at rest in this frame (the *proper* time). If $ds^2 = 0$ then the events can be connected by a photon, emitted at one event and received at the other, and for photons time stands still. If $ds^2 < 0$ then no observer can be present at both events, but $c(-ds^2)^{1/2}$ measures the *proper* distance (the distance as measured by radar methods) between the events. ds will be an invariant for observers with arbitrary (accelerated) motion, provided we assume the acceleration principle mentioned above.

The principle of equivalence, which forms the basis of the general theory of relativity, is expressed by saying that in the local freely falling frame special relativity holds and the expression for ds can be written in the form (4.14).

But this elimination of the effect of gravity by choosing a freely falling frame only works locally. For bodies far enough away from us we notice that they are falling to the centre of the Earth in a slightly different direction to ourselves. In general we will have to deal with *curved space–time* and the expression for the interval is

$$\begin{aligned} ds^2 = {} & g_{11}dx_1^2 + g_{22}dx_2^2 + g_{33}dx_3^2 + g_{44}dx_4^2 + 2g_{23}dx_2dx_3 \\ & + 2g_{24}dx_2dx_4 + 2g_{34}dx_3dx_4 + 2g_{12}dx_1dx_2 + 2g_{13}dx_1dx_3 \\ & + 2g_{14}dx_1dx_4, \qquad \text{where } x_4 = t, \end{aligned}$$

or, more compactly,

$$ds^2 = \sum_{\lambda,\mu=1}^{4} g_{\lambda\mu}\, dx_\lambda\, dx_\mu, \tag{4.15}$$

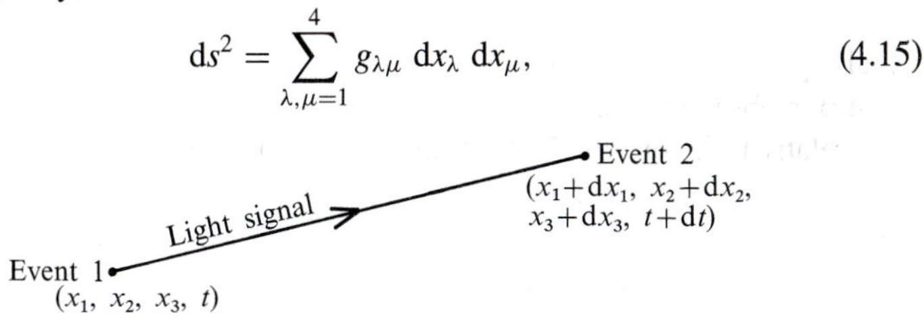

Fig. 4.4 Two neighbouring events connected by a light signal.

where the $g_{\lambda\mu}$ are functions of position and time which determine the curvature of space–time and hence the gravitational field. Eqn (4.15) describes the *metric* of space–time, and the $g_{\lambda\mu}$ are called the components of the *metric tensor* ($g_{\lambda\mu}$ is chosen equal to $g_{\mu\lambda}$).

In a normal three-dimensional Euclidean space, the metric is simply

$$ds^2 = dx_1^2 + dx_2^2 + dx_3^2$$

and so $g_{11} = g_{22} = g_{33} = 1$, and $g_{\lambda\mu} = 0$, $\lambda \neq \mu$. The metric of special relativity (called the Minkowski metric) is, by eqn (4.14),

$$ds^2 = -\frac{1}{c^2}(dx_1^2 + dx_2^2 + dx_3^2) + dx_4^2,$$

where $x_4 = t$, so

$$g_{11} = g_2 = g_{33} = -\frac{1}{c^2}; \quad g_{44} = 1, \quad g_{\lambda\mu} = 0, \quad \lambda \neq \mu.$$

But in a general frame of reference the $g_{\lambda\mu}$ will vary with position and time, and the geometry will be that of a curved space-time. This is the mathematical way of describing the effect of gravity.

In Newtonian theory and special relativity, the orbit of a free particle or a photon was a straight line. In the general theory the role of straight lines is taken by *geodesics*, the shortest routes between pairs of points in a curved space. It is then found that light passing near the Sun is deflected through a small angle of order $4GM_\odot/R_\odot c^2$ (in radians) where $M_\odot$ and $R_\odot$ are the mass and radius of the Sun. Suppose now the Sun were compressed by a very large factor until this quantity became greater than unity. Light near the Sun would then be so strongly deflected that in fact none could escape. The Sun would disappear from view, becoming a *black hole* (see p. 25). For massive stars it seems that this is indeed their probable fate (Section 2.4). Other effects of general relativity include the gravitational redshift, mentioned in Section 3.3, the time delay for radar signals reflected off the planets and passing near the Sun, and the advance of the perihelia of the planets, most pronounced for Mercury. These effects have all been measured and the predicted differences between general relativity and Newtonian theory are confirmed within the solar system to an accuracy of better than 1 per cent.

One of the most interesting applications of the bending of light is the *gravitational lens*. Light from a distant source is bent slightly round an intervening star, galaxy, or cluster of galaxies, which then act as a lens, distorting and amplifying the image of the distant source (Fig. 4.5). If the lens and the background source are perfectly aligned, then the image forms a ring of radius (the *Einstein* radius):

$$r_E = \{4GMDx(1-x)\}^{1/2}/c, \tag{4.16}$$

where M, d are the mass and distance of the lensing object, D is the distance of the background source, and $x = d/D$. If the alignment is not perfect then the ring will break up into two or more images.

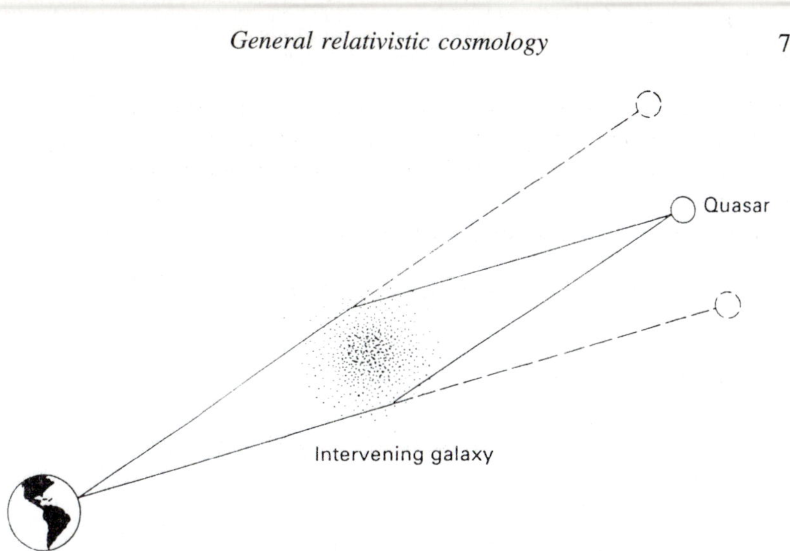

Fig. 4.5 Gravitational lensing of light from a distant quasar by an intervening galaxy. The light is bent slightly as it passes around the galaxy, generating two or more images of the quasar.

Gravitational lensing by low-mass stars in the halo of our Galaxy has been detected by the amplification effect of the lensing (*microlensing*). Many examples of lensing of distant ($z > 2$) galaxies and quasars by intervening galaxies have been found. Images of many rich clusters of galaxies show a pattern of circular arcs due to lensing of background galaxies by the cluster.

The binary pulsar, discovered in 1975, has provided one of the most impressive confirmations of general relativity. The system consists of two neutron stars in such close orbit around each other that significant emission of gravitational radiation occurs. This can be inferred from the gradual lengthening of the orbital period of the system.

4.5 General relativistic cosmology

It can be shown that the most general metric satisfying the cosmological principle is the *Robertson–Walker metric*:

$$ds^2 = dt^2 - \frac{R^2(t)}{c^2}\left(\frac{dr^2}{1 - kr^2} + r^2 d\theta^2 + r^2 \sin^2\theta \, d\phi^2\right), \qquad (4.17)$$

where (r, θ, ϕ) are spherical polar coordinates, and r is chosen for simplicity to be a *co-moving* radial coordinate, i.e. we are picking a coordinate system in which fundamental observers (see Section 4.1—we normally identify these with galaxies) have the same radial coordinate all the time, even if the universe expands or contracts. k is the curvature constant and if $k \neq 0$, it is convenient to redefine the units of the variable r so that $k = \pm 1$. $R(t)$ is again the *scale factor* and we still have

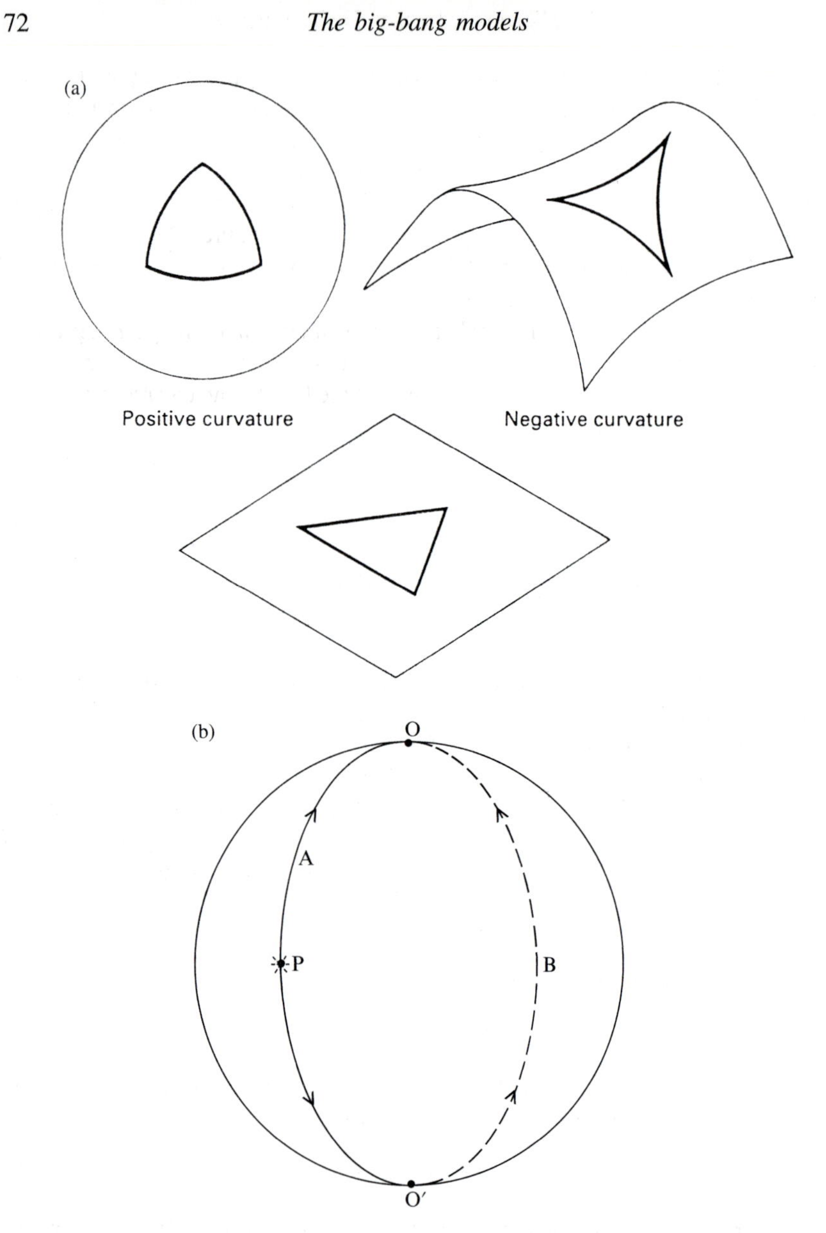

Fig. 4.6 (a) Illustration of spaces of positive ($k = 1$), zero ($k = 0$), and negative ($k = -1$) curvature. (b) An example of a two-dimensional space of positive curvature, the surface of a sphere. A signal confined to the surface and travelling by the shortest route will travel on a geodesic, in this case the great circle joining the source P and observer O. the signal has two alternative routes, PAO and PO′BO.

the result that, as t changes, all spatial dimensions are simply scaled up by the factor $R(t)$. Thus exactly the same motions, isotropic expansion or contraction, are permitted in general relativity as in Newtonian cosmology (Section 4.3).

If we consider all events with the same value of t, so that $\mathrm{d}t = 0$, the metric (4.17) can be shown to correspond to a three-dimensional space of constant curvature. If $k = +1$, we have a space of positive curvature (elliptic space—see Fig. 4.6); if $k = 0$ we have normal flat space; and if $k = -1$, we have a space of negative curvature (hyperbolic space).

To solve for $R(t)$ we must substitute this metric into the *field equations*, differential equations relating the metric functions $g_{\lambda\mu}$ to the density and pressure of matter—the general relativistic analogue of the Newtonian equations of motion for a fluid. When we do this we again obtain not only eqn (4.10), which is not too surprising since it expresses the conservation of mass, but also eqns (4.12) and (4.13).

At first sight it seems amazing that the general theory of relativity yields exactly the same cosmological models as Newtonian theory (the two theories do not yield the same results for the motion of the planets). But it is less strange when we remember (a) that the general theory of relativity is designed to reduce to Newton's law of gravitation when the gravitational field is very weak, e.g. in our cosmological neighbourhood, and (b) the cosmological principle requires that each neighbourhood be identical to every other.

Before we study the properties of eqn (4.13), we should note that in the original form of the field equations proposed by Einstein, an additional term appeared, the so-called 'cosmological term'. Later, Einstein argued that this term should be dropped, and this is the view of most relativists today. In Chapter 8 certain interesting consequences of the cosmological term are discussed, but for the moment it is set equal to zero.

4.6 Classification of cosmological models

The Milne model, $\rho = 0, k = -1$

This is a universe of particles of negligible mass, so can also be called the special relativity cosmology. The solution of eqn (4.13) is

$$R(t) = \pm ct, \tag{4.18}$$

where we have chosen $t = 0$ to correspond to $R = 0$. The universe expands (or contracts) uniformly and monotonically (Fig. 4.7).

The Einstein–de Sitter model, $k = 0$

The solution of eqn (4.13) is

$$R(t) = \pm R_0 (t/t_0)^{2/3} \tag{4.19}$$

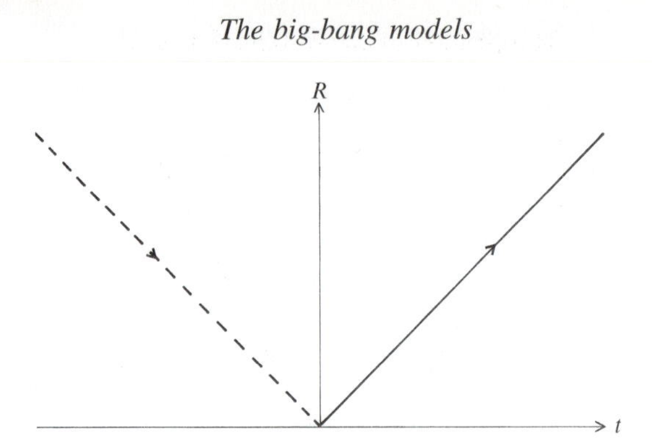

Fig. 4.7 The Milne (special relativity) model $R \propto t$. The contracting solution is shown as a broken line.

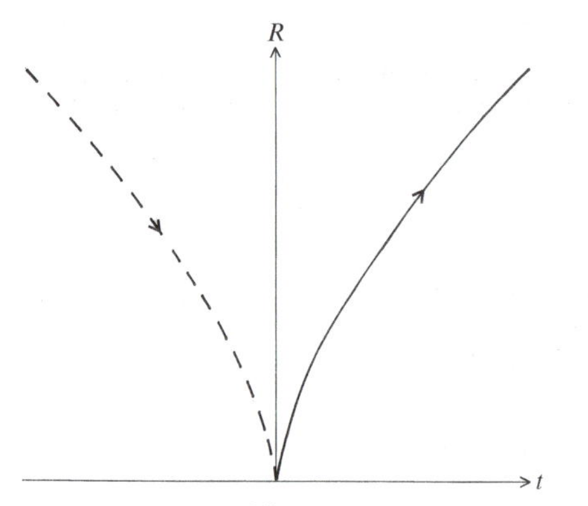

Fig. 4.8 The Einstein–de Sitter model $R \propto t^{2/3}$. The contracting solution is shown as a broken line.

and the universe again expands monotonically, but at an ever-decreasing rate (see Fig. 4.8).

$\rho > 0, k = -1$

$\dot{R}^2 > 0$ for all R, so R keeps changing monotonically (Fig. 4.9). As t gets very large, $\dot{R} \to \pm c$, so the universe looks more and more like a Milne model. As the galaxies get very far apart, their mutual gravitational attraction plays an ever weaker role in determining their motion.

$k = +1$

$\dot{R}^2 = 0$ for a particular value of R,

$$R_\text{c} = \frac{8\pi G \rho_0 R_0^3}{3c^2}$$

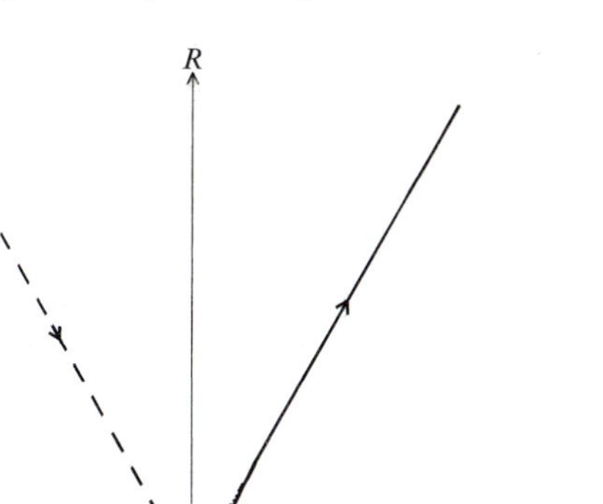

Fig. 4.9 The general $k = -1$ case. From small t, $R \propto t^{2/3}$, and for large t, $R \propto t$ (asymptote shown dotted). The universe expands (or contracts—broken line) monotonically.

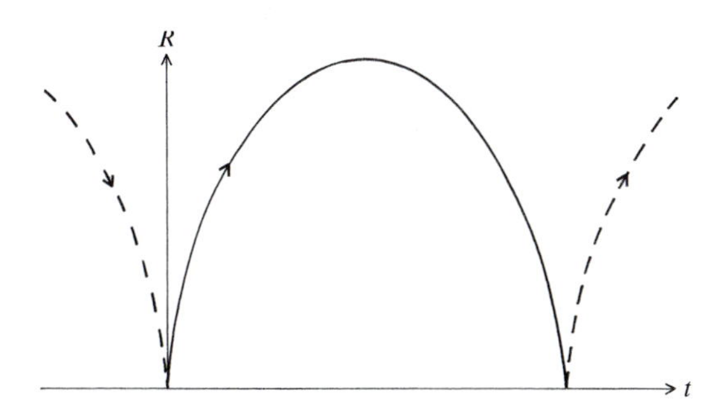

Fig. 4.10 The general $k = +1$ case. An 'oscillating' universe. It is not known if the universe can go through more than one cycle.

and since $\ddot{R} < 0$ for all R, the expansion is halted (by the mutual gravitational attraction between the galaxies) and turned into a contraction. We have an 'oscillating' universe (see Fig. 4.10).

Note that (i) whatever the value of k, the first term on the right-hand side of eqn (4.13) must dominate for small values of R, so the universe always looks like an Einstein–de Sitter model in the early stage, if $\rho > 0$, (ii) all these models are big-bang universes, in that $R \to 0$ at a finite time in the past (for expanding models), and the universe emerged from a 'singularity' (infinite density) at that time.

This follows very simply from eqn (4.12), which tells us that the $R(t)$ curve must be concave downwards and so must intersect the $R = 0$ axis at a finite time in the past. Note that this big bang is not an explosion into a pre-existing void. Because of the assumption of homogeneity, the whole universe in involved in the expansion and there is no 'outside' to expand into.

4.7 Cosmological parameters

We have already encountered (Section 3.3) the *Hubble parameter*

$$H(t) = \dot{R}(t)/R(t). \tag{4.20}$$

We now define the *deceleration parameter*

$$q(t) = -R(t)\ddot{R}(t)/\dot{R}^2(t) \tag{4.21}$$

and the *density parameter*

$$\Omega(t) = 8\pi G\rho(t)/3H^2(t). \tag{4.22}$$

Eqn (4.12) implies that $\Omega(t) = 2q(t)$ at all values of t and (4.13) implies that

$$kc^2 = R^2H^2(\Omega - 1). \tag{4.23}$$

Thus whether the curvature constant k is $+1$, 0, or -1 is determined by whether Ω is greater than, equals, or is less than, 1. The only models in which Ω (and hence q) does not change with time are the Milne ($\Omega = q = 0$) and Einstein–de Sitter ($\Omega = 2q = 1$) models.

The currently accepted values of these parameters at the present epoch $t = t_0$, are:

$$H_0 = 50 - 100\,\mathrm{km\,s^{-1}Mpc^{-1}}$$

so

$$\tau_0 = H_0^{-1} = 1.5 \pm 0.5 \times 10^{10} \text{ years}$$

$-1 \leq q_0 \leq 2$ (from the magnitude–redshift test—see Chapter 7)

$0.03 \leq \Omega_0 \leq 1$ (from the abundances of the light elements—see Chapter 5—and from dynamical arguments—see Chapter 6).

The value of the density appropriate to the Einstein–de Sitter model would be

$$\rho_{\mathrm{ES}} = 3H_0^2/8\pi G = 5 \times 10^{-27}(H_0/50\ \mathrm{km\ s^{-1}\ Mpc^{-1}})^2\ \mathrm{kg\ m^{-3}}. \tag{4.24}$$

This is often called the *critical density*, since if $\rho_0 > \rho_{\mathrm{ES}}$ we are in an oscillatory universe and if $\rho_0 < \rho_{\mathrm{ES}}$ we are in a monotonic expanding one.

4.8 The age of the universe

We can write

$$t_0 = \int_0^{t_0} \mathrm{d}t = \int_0^{R_0} \frac{\mathrm{d}R}{\dot{R}}$$

and using eqns (4.13) and (4.20)–(4.23):

$$t_0 = \tau_0 \int_0^1 \frac{\mathrm{d}x}{(2q_0/x + 1 - 2q_0)^{1/2}} \tag{4.25}$$

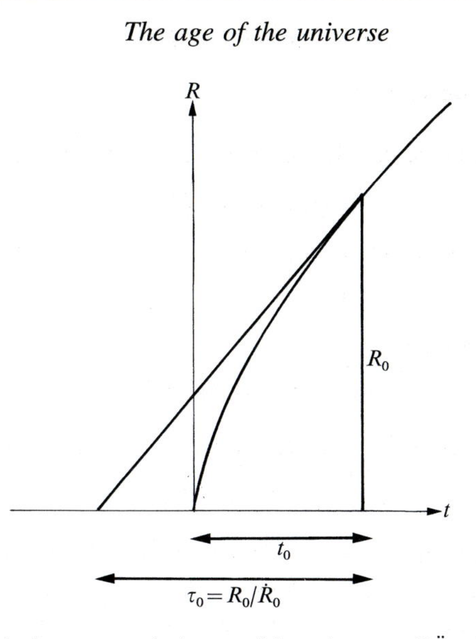

Fig. 4.11 The Hubble time τ_0 exceeds the age of the universe t_0, if $\ddot{R} \leq 0$ for all t.

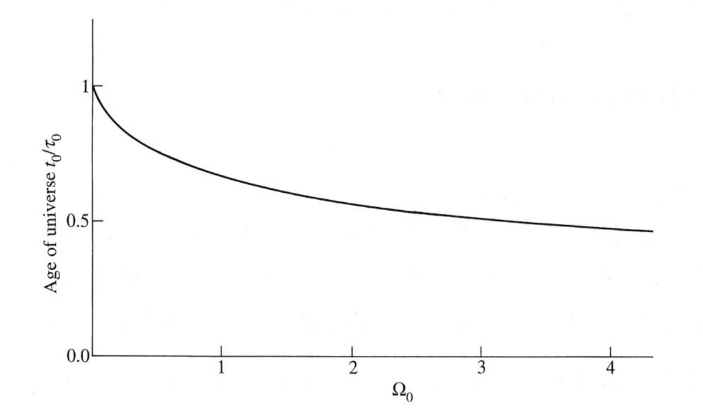

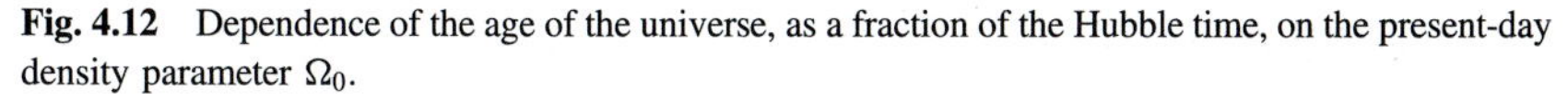

Fig. 4.12 Dependence of the age of the universe, as a fraction of the Hubble time, on the present-day density parameter Ω_0.

where we have written x for R/R_0. If $q_0 = 0$, $t_0 = \tau_0$; if $q_0 = 1/2$, $t_0 = 2\tau_0/3$.

For $0 < q_0 < 1/2$, and $q_0 > 1/2$, the substitutions

$$x = \frac{2q_0}{1 - 2q_0} \sinh^2 \theta, \text{ and } x = \frac{2q_0}{2q_0 - 1} \sin^2 \theta \tag{4.26}$$

respectively, allow eqn (4.25) to be evaluated. The age of the universe is less than or equal to the Hubble time in all models: the reason for this is illustrated in Fig. 4.11. Thus if $H_0 = 100$ km s^{-1} Mpc^{-1}, the age of the universe is less than or equal to 9.8×10^9 years. This is somewhat less than the best current estimates of the age of our Galaxy (Section 3.7). For this reason the value $H_0 = 50$ km s^{-1} Mpc^{-1} is adopted in the remainder of this book.

4.9 Horizon

Even if we live in an open universe with infinitely many galaxies in it, the light from only a finite number of them will have reached us so far. There is therefore a (particle) *horizon* which divides those particles in the universe that we can already have observed from those that we cannot yet know anything about. As time proceeds, galaxies swim into view. At first they are seen with very large redshift. Since the expansion of the universe is slowing down ($q > 0$), the redshift of any particular galaxy decreases with time.

The radius of the horizon at time t is of order ct so it encompassed a much smaller volume of matter at early times. It is thus something of a paradox that when we look at the microwave background in two opposite directions, conditions are identical to one part in 10^5 despite the two regions never having yet been in causal contact. This is often called the *horizon problem* (see Section 5.4).

4.10 Problems

4.1 Show by means of the substitutions (4.26) in (4.25) that the age of the universe satisfies

$$t_0/\tau_0 = \Omega_0 \frac{\left[\cos^{-1}\left(\frac{(2-\Omega_0)}{\Omega_0}\right) - \frac{2(\Omega_0-1)^{1/2}}{\Omega_0}\right]}{2(\Omega_0 - 1)^{3/2}}, \qquad \text{if } k = 1,$$

$$= \frac{2}{3}, \qquad \text{if } k = 0,$$

$$= \frac{\Omega_0\left[\frac{2(1-\Omega_0)^{1/2}}{\Omega_0} - \ln\left(\frac{2-\Omega_0+2(1-\Omega_0)^{1/2}}{\Omega_0}\right)\right]}{2(1 - \Omega_0)^{3/2}}, \qquad \text{if } k = -1,$$

Find the age of the universe if $H_0 = 66$ km s^{-1} Mpc^{-1} and $\Omega_0 = 0.1$. If $t = 1.3 \times 10^{10}$ years, and $0.1 \le \Omega_0 \le 1$, what is the permissible range of H_0?

4.2 Show that q is constant in the Milne and Einstein–de Sitter models.

4.3 Show that the Friedmann eqn (eqn (4.13)) has the solution

$$R(t) = \frac{c\tau_0\Omega_0(1 - \cos 2\Psi)}{2(\Omega_0 - 1)^{3/2}},$$

$$t = \frac{\Omega_0\tau_0(2\Psi - \sin 2\Psi)}{2(\Omega_0 - 1)^{3/2}}, \qquad \text{if } k = 1$$

and

$$R(t) = \frac{c\tau_0\Omega_0(\cosh 2\Psi - 1)}{2(1 - \Omega_0)^{3/2}},$$

$$t = \frac{\Omega_0\tau_0(\sinh 2\Psi - 2\Psi)}{2(1 - \Omega_0)^{3/2}}, \qquad \text{if } k = -1,$$

where ψ ranges over all real values.

5
Early stages of the big bang

5.1 Universe with matter and radiation

So far we have constructed models of universes filled only with matter. Although radiation contributes less than 1 per cent of the average energy per unit volume in the universe at the present epoch, it plays a crucial and dominant role in the early stages.

We can study the evolution of a universe containing matter and radiation by applying the first law of thermodynamics to an element of the substratum. This gives the same answer as applying the full field equations of general relativity.

The first law of thermodynamics states that the change in energy of an expanding system equals the work done by the pressure

$$dE = -p\,dV, \tag{5.1}$$

where E, p, V are the energy, pressure, and volume of the element.

We now use Einstein's equation

$$E = Mc^2 \tag{5.2}$$

where M includes both the contribution of the matter and the mass equivalent of the radiant energy, i.e.

$$E = (\rho_m + \rho_r)Vc^2 = \rho Vc^2 \tag{5.3}$$

where ρ_m is the density of the matter, ρ_r is the mass density of the radiation ($= U/c^2$, where U is the energy density), and ρ is the total mass density of matter and radiation. Since the volume of an element of the substratum is proportional to $R^3(t)$ we have, from eqn (5.1),

$$\frac{d}{dt}(\rho R^3) + \frac{p}{c^2}\frac{d}{dt}(R^3) = 0 \tag{5.4}$$

This is the fundamental equation relating the density and pressure in a universe containing matter and radiation.

Universe with matter only

If the pressure of the matter can be neglected, eqn (5.4) reduces to

$$\frac{d}{dt}(\rho_m R^3) = 0$$

and so $\rho_m \propto R^{-3}$ as before (eqn (4.10)).

Universe with radiation only

It can be shown that the relation between the pressure and density of radiation is

$$p_{\mathrm{r}} = \rho_{\mathrm{r}} c^2/3 \tag{5.5}$$

Thus from eqn (5.4),

$$\frac{\mathrm{d}}{\mathrm{d}t}(\rho_{\mathrm{r}} R^3) + \frac{1}{3}\rho_{\mathrm{r}}\frac{\mathrm{d}}{\mathrm{d}t}(R^3) = 0,$$

or

$$3\rho_{\mathrm{r}} R^2 \dot{R} + \dot{\rho}_{\mathrm{r}} R^3 + \rho_{\mathrm{r}} R^2 \dot{R} = 0$$

so

$$\frac{\mathrm{d}}{\mathrm{d}t}(\rho_{\mathrm{r}} R^4) = 0. \tag{5.6}$$

This integrates to

$$\rho_{\mathrm{r}} = \rho_{\mathrm{r},0}\Big(R(t)/R_0\Big)^{-4}. \tag{5.7}$$

When this is substituted in eqn (4.12), assuming $k = 0$, we find $\dot{R}^2 \propto R^{-2}$, $\dot{R} \propto R^{-1}$, which integrates to

$$R(t) \propto t^{1/2}. \tag{5.8}$$

This characterizes the motion of a radiation-dominated universe in its early stages, since the term in eqn (4.13) in kc^2 becomes negligible if R is sufficiently small.

Universe containing matter and radiation

If we neglect the contribution of matter to the pressure, so that $p = p_r$ then eqn (5.4) becomes

$$\frac{\mathrm{d}}{\mathrm{d}t}(\rho_{\mathrm{m}} R^3) + \frac{1}{R}\frac{\mathrm{d}}{\mathrm{d}t}(\rho_{\mathrm{r}} R^4) = 0. \tag{5.9}$$

If we assume strict conservation of matter, i.e. we neglect any conversion of matter to radiation, then each of the two terms in eqn (5.9) will be separately zero:

$$\frac{\mathrm{d}}{\mathrm{d}t}(\rho_m R^3) = 0, \quad \frac{\mathrm{d}}{\mathrm{d}t}(\rho_r R^4) = 0$$

and so

$$\rho_{\mathrm{m}} = \rho_{\mathrm{m},0}(R/R_0)^{-3},\ \rho_r = \rho_{\mathrm{r},0}(R/R_0)^{-4}. \tag{5.10}$$

However small the current value of the ratio of the density of radiation to matter $\rho_{r,0}/\rho_{m,0}$ (current value $\sim 10^{-3}$), there was an epoch in the past, given by

$$R_{crit} = \frac{\rho_{r,0}}{\rho_{m,0}} R_0 \tag{5.11}$$

such that $\rho_r > \rho_m$ for $R < R_{crit}$, and $\rho_r \gg \rho_m$ for $R \ll R_{crit}$. Radiation would have been the dominant form of energy at early epochs, and eqn (5.8) would have been valid.

We call epochs such that $\rho_r > \rho_m$ the *radiation-dominated era* and epochs such that $\rho_m > \rho_r$ *the matter-dominated* era.

5.2 The fireball

At the present epoch radiation traverses the universe freely, with only a small probability of being scattered by gas or dust. In addition to occurring in galaxies, such material may be spread more or less uniformly throughout space, but so tenuously that a photon is likely to travel several Hubble distances (see p. 50) before being scattered or absorbed. We say that the universe is transparent or *optically thin* at the present epoch. It is easy to see that this was not always so. The typical size of a galaxy is 10 kpc, while the average spacing between galaxies is of order 1 Mpc. If we run the universe backwards through a change Z in the scale factor of 100, without galaxies altering, i.e.

$$Z = R(t_0)/R(t) = 100,$$

then the galaxies would all have been touching. This makes it likely that the break-up into galaxies occurred during epochs such that $100 \geq Z > 1$. Of course, galaxies would not remain unaltered during this 'rewind' of the universe. We would see the dust falling back into the stars from which it has blown out, and the stars dissolving into the gas clouds from which they formed. The atoms from which our complex chemistry derives would have broken down into hydrogen with an admixture of helium. We do not have any direct evidence on how smooth the universe was on the scale of galactic masses at epochs such that $Z > 100$, but it is natural to consider a picture of a fairly uniformly distributed gas of hydrogen and helium, spotted with the density irregularities which are to condense into the galaxies. We shall discuss the problem of how galaxies may have formed in the next chapter.

All the while that we are running the universe backwards, the energy density, and hence the temperature, of the radiation is building up, by eqn (5.10). Sooner or later the gas will start to be significantly heated by the radiation. The crucial moment comes when the temperature of the matter reaches about 3000 K, for then the hydrogen starts to become ionized. This brings into play the enormous scattering power of free electrons and puts an end to the transparency of the universe to radiation. We call this moment the epoch of *decoupling* of radiation and matter. Prior to this, they are locked together in thermal equilibrium. This means that the radiation has the Planck blackbody spectrum (Section 1.4), so the total energy density of the radiation is

$$\rho_r c^2 = \int_0^\infty u_r(\nu) c^2 \, d\nu,$$

where the specific energy density is $u_r(\nu) = 4\pi I(\nu)/c$, and $I(\nu)$ is the intensity (eqn 1.2). Thus

$$\rho_r c^2 \propto \int_0^\infty \frac{\nu^3 d\nu}{\exp(h\nu/kT_r) - 1} \propto T_r^4, \tag{5.12}$$

where T_r is the radiation temperature, and then eqn (5.10) implies that

$$T_r \propto 1/R(t). \tag{5.13}$$

As $R \to 0$, $T_r \to \infty$, which explains the use of the term *fireball* to describe this optically thick phase of a big-bang universe. Since there is thermal equilibrium prior to the epoch of decoupling, the temperature of the matter is the same as that of the radiation:

$$T_m = T_r \,. \tag{5.14}$$

What happens after the epoch of decoupling? Certainly eqn (5.14) will no longer hold and the matter will cool off rapidly. However, it turns out that the effect of the expansion of the universe on the radiation is to preserve its blackbody spectrum, with the radiation temperature continuing to fall according to eqn (5.13) (see Ex. 5.1). This provides the most natural explanation of the 2.7 K blackbody microwave background radiation described in Section 1.8. A more detailed summary of the observations is shown in Fig. 5.1. The discovery of this radiation provided the most spectacular confirmation to date of the hot big-bang picture of the universe. The COBE satellite, launched in 1989, provided a superb confirmation of the blackbody nature of the microwave background spectrum (Fig. 5.1b).

Now, from eqn (5.13) we can identify the epoch of decoupling as

$$Z = R(t_0)/R(t) = 3000/2.7 \sim 1000.$$

The density of matter at this time would have been $\rho_m \sim (10^3)^3 \rho_{m,0} \sim 10^{-18}$ kg m^{-3} (assuming $\rho_{m,0} \sim 10^{-27}$ kg m^{-3}), about 1000 times the average density of matter inside our Galaxy.

Coincidentally, the critical epoch separating the radiation-dominated from the matter-dominated eras (Section 5.1) is also about $Z \sim 1000$ for $\rho_{m,0} \sim 10^{-27}$ kg m^{-3}. Hence for $Z > 1000$ we can use $R(t) \propto t^{1/2}$ (eqn (5.8)), and for $1000 > Z \gg 1$ we can use $R(t) \propto t^{2/3}$ (eqn (4.19)). In Figs 5.2 and 5.3, which show the variation of the energy densities and temperatures of matter and radiation with epoch, we use the latter right up to $Z = 1$, for illustration.

Fig. 5.1 (a) The Bell Laboratories' antenna with which Penzias and Wilson discovered the microwave background radiation.

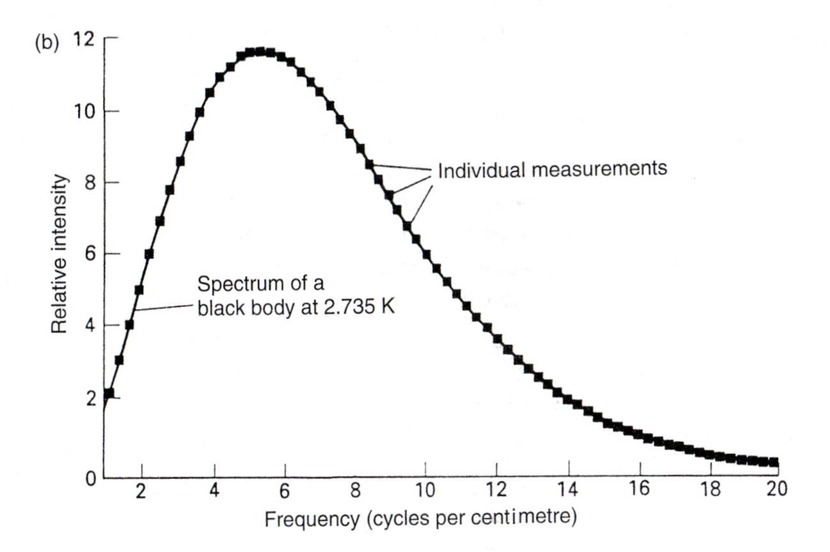

(b) The spectrum of the microwave background radiation measured by COBE, compared with a $T = 2.735$ K black body.

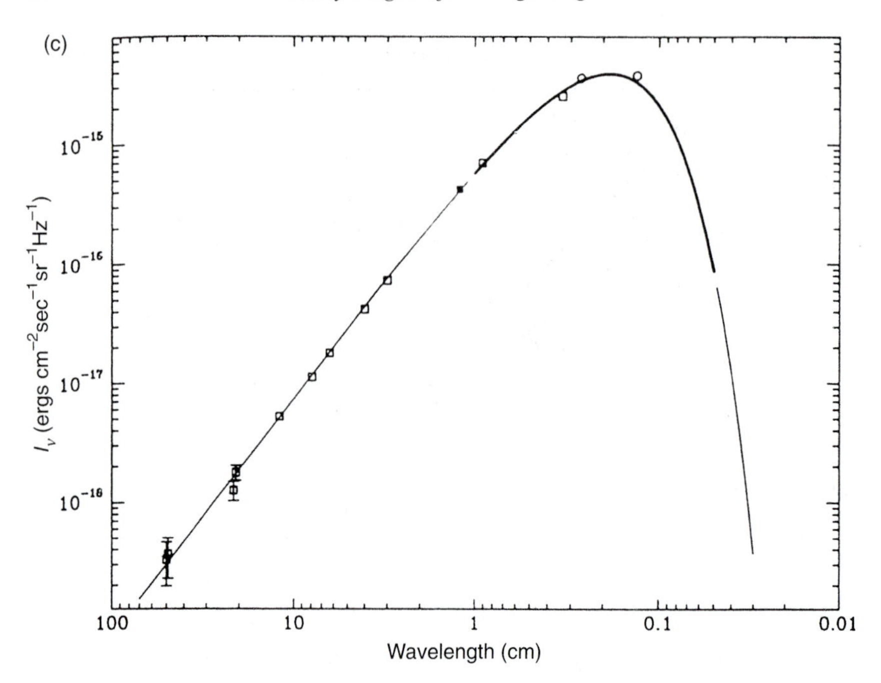

Fig 5.1 (Cont.) (c) Spectrum of the cosmic background radiation (Wilkinson 1992). The scale is chosen to emphasize the measurements in the long wavelength Rayleigh–Jeans part of the spectrum. The COBE measurement from Fig. 5.1(b) is shown as the solid curve near the peak of the spectrum, the more precise of the Dicke radiometer measurements are plotted as the squares, and the excitation temperatures of the first two excited levels of the interstellar molecule cyanogen as circles. The thin line is the Planck blackbody spectrum. Where error flags are not shown they are comparable to or smaller than the sizes of the line or symbol. The measurements at long and short wavelengths are limited by emission from the galaxy, which is larger than the extragalactic part at wavelengths longer than 30 cm and shorter than 400 μm.

5.3 Helium production

When the temperature was between 10^{10} K and 10^9 K, some proportion of the hydrogen was converted by thermonuclear fusion into helium, the exact amount depending on the density of the matter during this phase. For $\rho_{\mathrm{m},0} \sim 10^{-27}$ kg m^{-3},

$$\rho_{\mathrm{m}}(T = 10^{10}\,\mathrm{K}) \sim 10^{+1.8}\,\mathrm{kg\,m^{-3}},$$

and it is found that the fraction of matter converted to helium would be almost exactly the 24 per cent by mass that we need to explain the composition of our Galaxy (Fig. 5.4 and Table 5.1). This provides a second major success for big-bang cosmology.

We have pushed the fireball scenario back to within 1 second of the big bang. Extrapolating what is known about elementary particles, we can get even closer,

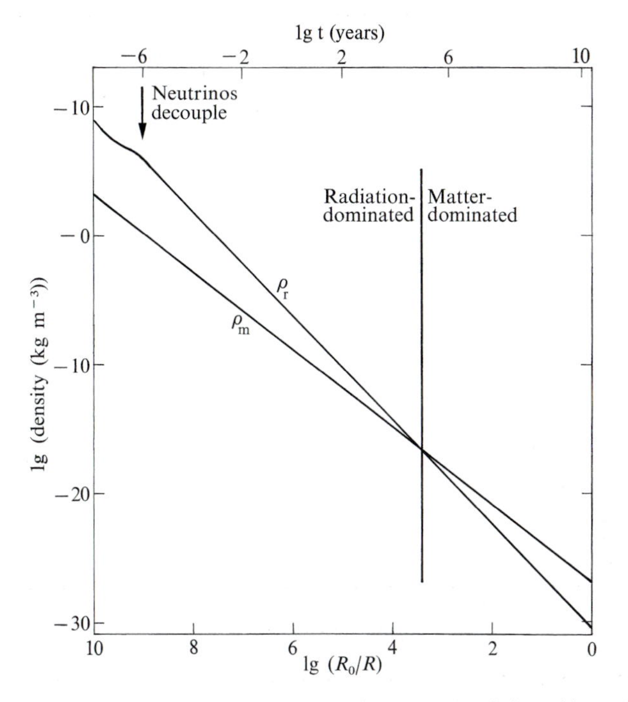

Fig. 5.2 The variation of the density of matter and radiation with epoch.

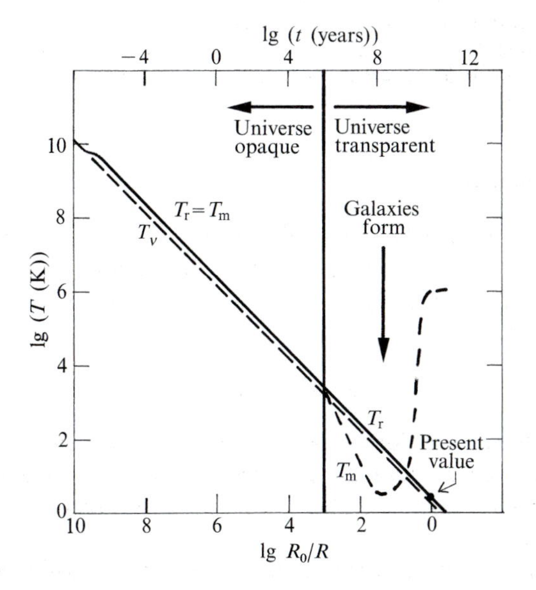

Fig. 5.3 The variation of the temperature of matter and radiation with epoch.

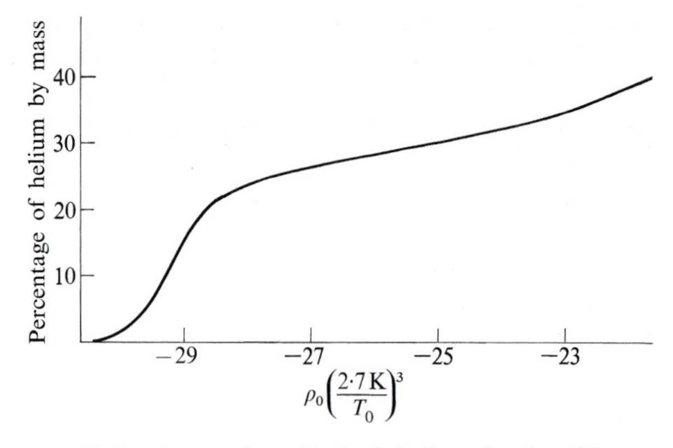

Fig. 5.4 Percentage of helium by mass formed in the fireball as a function of the present density of the universe.

but we should bear in mind that even the most optimistic interpretation of the 0.001 per cent isotropy of the blackbody radiation guarantees our simple isotropic models only back to $Z \sim 10^8$, so we are treading on thin ice.

Let us now look at the evolution of the universe during the fireball phase in more detail. For any particular elementary particle of mass m, there is an epoch at which the temperature is such that $kT \sim mc^2$. Prior to this epoch a collision between two photons can result in the creation of a particle pair, the particle and its antiparticle. The thermal equilibrium between matter and radiation ensures that when $kT > mc^2$ there are roughly as many particles of this type as photons. When the temperature drops below mc^2/k, particle pairs can no longer be created and the particles and antiparticles rapidly annihilate, provided they are abundant enough to collide frequently.

The elementary particles can be subdivided into the lighter particles which do not take part in strong nuclear interactions, or *leptons* (neutrinos, electrons, muons, and tauons and their antiparticles), and the *hadrons*, which do. Hadrons can be further subdivided into *baryons* (protons, neutrons, and the unstable hyperons) and *mesons*. The critical temperatures for the hadrons are $> 10^{12}$ K and this phase of the universe is known as the hadron era. What the hadron era was like depends on what is the correct theory for strong interactions. There would have been a great variety of particles in thermal equilibrium with each other, including photons, leptons, and, possibly, mesons and nucleons (i.e. neutrons and protons) and their antiparticles. According to the quark theory of strong interactions the hadrons would then have been broken down into free quarks.

The subsequent history of the fireball can be divided into four regimes:

(a) $T \sim 10^{12}$ K ($t \sim 10^{-4}$ s) muons annihilate; muon neutrinos and antineutrinos decouple from everything else.

(b) $T < 10^{11}$ K ($t > 0.01$ s): the neutron–proton mass difference (1.3 MeV, corresponding to $T = 1.5 \times 10^{10}$ K) begins to shift the small nucleonic

Table 5.1 Relative cosmic abundances of the most common elements

(a) Helium

Location; method of evaluation	Percentage helium by mass
Initial value in Sun, from stellar evolution calculations	22–27
Solar cosmic rays	20–26
Globular cluster and other old stars (apart from some anomalous stars with no visible helium)	20–30
Planetary nebulae	40
H_{II} regions	27
Blue compact galaxies	24
Best primordial value	24

(b) The most common elements (abundance by mass relative to hydrogen)

Hydrogen	1
Helium	0.32
Carbon	4.0×10^{-3}
Nitrogen	1.3×10^{-3}
Oxygen	1.05×10^{-2}
Neon	1.7×10^{-3}
Magnesium	6.4×10^{-4}
Silicon	9.4×10^{-4}
Sulphur	5.1×10^{-4}
Argon	2.5×10^{-4}
Iron	2.25×10^{-3}
Nickel	1.2×10^{-4}

(c) Deuterium

	(D/H)
Sun	$3 \pm 1 \times 10^{-5}$
Interstellar gas	$1.8 \pm 0.4 \times 10^{-5}$
Dense molecular clouds	$2 - 5 \times 10^{-3}$ (due to chemical fractionation)

contamination towards more protons and fewer neutrons, through equilibrium of the weak interaction processes.

$$\mathrm{n} + \nu_e \longleftrightarrow \mathrm{p} + \mathrm{e}^-, \quad \mathrm{n} + \mathrm{e}^+ \longleftrightarrow \mathrm{p} + \bar{\nu}_e$$

$$\mathrm{n} \longleftrightarrow \mathrm{p} + \mathrm{e}^- + \nu_e$$

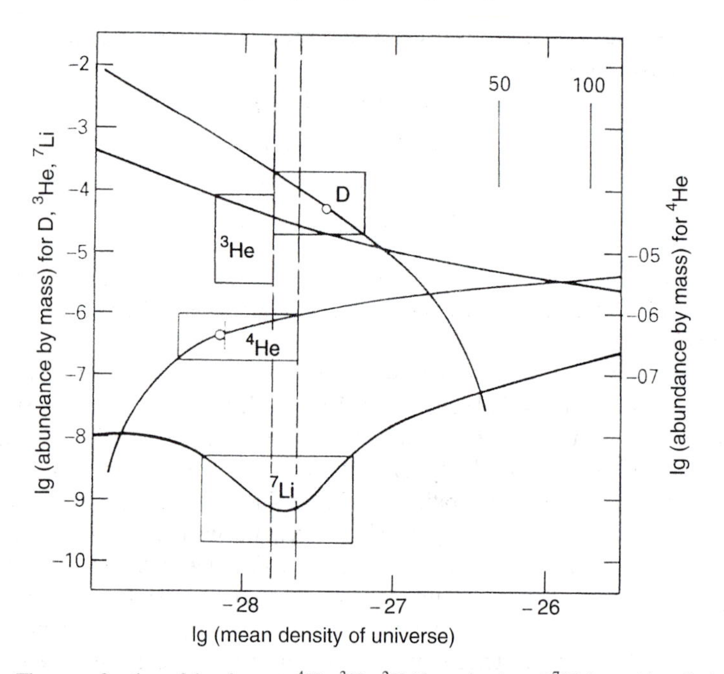

Fig. 5.5 The mass fraction of the elements ^{4}He, ^{3}He, ^{2}H (deuterium), and ^{7}Li formed in a fireball as a function of the present density of the universe, compared with observed primordial values (boxes). A density not far from 2×10^{-28} kg m^{-3} seems to be required. The corresponding values for the critical density of two values of the Hubble constant are indicated by vertical lines.

The neutron–proton equilibrium ratio is determined by the temperature

$$N_n/N_p = \exp(-1.5 \times 10^{10}/T). \tag{5.15}$$

At $T \sim 10^{10}$ K ($t \sim 1$ s) electron neutrinos and antineutrinos start to decouple from everything else.

(c) $T \sim 5 \times 10^9$ K ($t \sim 4s$): electrons and positrons annihilate with each other and this, together with the cooling of the neutrinos by the expansion of the universe, leads to the virtual 'freezing-out' of the neutron–proton ratio. The only remaining process is neutron decay ($n \to p + e^- + \overline{\nu}_e$, β-decay) and the resulting fraction of neutrons to all nucleons is

$$X_n = N_n/(N_n + N_p) \approx 0.16 \exp(-t/1013\,\text{s}). \tag{5.16}$$

The annihilation of the electrons and positrons raises the photon temperature to 1.4 times that of the neutrinos (see Fig. 5.3).

(d) $T \sim 10^9$ K ($t \sim 10^2$ sec): nucleosynthesis begins, yielding ^{4}He and a trace of ^{2}H (deuterium), ^{3}He, ^{7}Li, and other elements (see Fig. 5.5). Number densities are too low to allow nuclei to be built up directly in many-body collisions like

$2n + 2p \rightarrow {}^4He$. Instead complex nuclei must be built up in sequences of two-body reactions like

$$n + p \longleftrightarrow {}^2H + \gamma$$
$${}^2H + {}^2H \longleftrightarrow {}^3He + n$$
$${}^3He + n \longleftrightarrow {}^3H + p$$
$${}^3H + {}^2H \longleftrightarrow {}^4He + n.$$

The crucial step is the formation of deuterium which, because of its low binding energy (2.2 MeV), is destroyed as soon as it is made until the temperature drops to $\sim 10^9$ K.

The absence of stable nuclides with atomic number $A = 5$ and 8 means that very little production of elements heavier than helium takes place. Almost all neutrons end up in ^{4}He nuclei, which have by far the highest binding energy of all nuclei with $A < 5$. Nucleosynthesis turns off the decay of free neutrons and fixes the neutron–proton ratio at the value just before the onset of nucleosynthesis. After nucleosynthesis is over, we have essentially nothing left but free protons and helium nuclei, so the mass-fraction in the form of helium is simply twice the fraction of neutrons to all nucleons just before the onset of nucleosynthesis, $2X_n$. This depends weakly on the density of matter in the universe and hence the time that the neutrons decay according to eqn (5.16). Deuterium, on the other hand, is very sensitive to ρ_m because its final abundance is controlled by the processes which destroy it, and the cross-section for these reactions is proportional to ρ_m^2. The observed primordial abundances of helium, deuterium, and lithium imply (see Fig. 5.5) that the density of the universe today in the form of baryons is given by

$$\Omega_{b,0} = 0.05 \pm 0.01(H_0/50)^{-2}. \tag{5.17}$$

Finally, when the temperature drops to 3000 K ($t \sim 3 \times 10^5$ years), the temperature of the matter becomes too low to keep hydrogen ionized. Protons and electrons combine to form neutral atoms of hydrogen. The matter therefore suddenly becomes transparent to radiation, and matter and radiation decouple. The process of 'recombination' is so rapid that there are no significant distortions of the blackbody spectrum during this phase and, as discussed in Section 5.2, the subsequent expansion of the universe preserves the blackbody form of the spectrum, while changing the temperature according to eqn (5.13).

5.4 The very early universe

In the previous section we discussed the evolution of the universe from a time 10^{-4} seconds after the big bang onwards, at which time the temperature is 10^{12} K. The physics of matter under these conditions is well understood from terrestrial accelerator experiments. Prior to this epoch we start extrapolating into the realm of less well-established physics.

According to the current view of the nature of hadrons, they are all made up of triplets of quarks, of which there are six different kinds or 'flavours': up and down, strange and charmed, top and bottom. The proton is composed of two up quarks and one down quark, while the neutron is composed of one up and two down quarks. The attractive force, which holds the quarks together, is provided by particles known as 'gluons', and the detailed theory of this force is known as quantum chromodynamics (QCD). The name arises because another property of quarks has been given the name 'colour' (a quark can exist in one of three possible colour states).

At $t \sim 10^{-6}$ s ($T \sim 10^{13}$ K), quarks and their antiparticles annihilate each other, and their residues combine to form protons and neutrons in equal numbers. Prior to this epoch there would have been a soup of quarks, leptons, and their antiparticles.

Pushing back further to $t \sim 10^{-12}$ s ($T \sim 10^{15}$ K), it is believed that the electromagnetic force and the weak nuclear force (the one responsible for β-decay) would have been combined into a single force, the *electroweak* force, prior to this epoch. At this epoch a phase transition occurs in which the electroweak force splits into the two forces that we know today. The theory of the electroweak unification, developed by Weinberg and Salam, has been confirmed in accelerator experiments at CERN by the detection of the W and Z particles hypothesized to be responsible for the force.

On much shakier ground we can extrapolate back a further 23 powers of ten in time to $t \sim 10^{-35}$ s, prior to which time it is proposed that the strong nuclear force (responsible for holding protons and neutrons together in the nuclei of atoms) and the electroweak force would have been unified into the grand unified force. Grand Unified Theories (GUTs) predict that the proton will decay on a time-scale of about 10^{31} years and sensitive experiments are under way to test this prediction.

In 1981 Alan Guth suggested that the phase transition associated with the break-up of the grand unified force could leave the universe (or our portion of it at least) in a state of 'false vacuum', in which the vacuum has a colossal energy-density associated with it. This vacuum energy-density acts like a cosmical repulsion (see Section 8.2) and the universe embarks on an exponential expansion which inflates the universe by the incredible factor 10^{28}, or perhaps even more, in a very brief instant of time. The inflationary period ends when the vacuum energy-density transforms into matter and radiation and the expansion of the universe continues as described above. *Inflation* permits a solution of the 'horizon' problem (Section 4.9) because regions of the universe seen in opposite directions on the sky through the microwave background, although not in communication with each other today, would have been in communication before the inflation began. The inflation also solves what is known as the 'flatness' problem. Today we know that the density parameter lies in the range $0.03 < \Omega_0 < 2$, which seems quite a broad range of possibilities. However at $t = 1$ s after the big bang the quantity $|1 - \Omega_0|$, which measures how much the

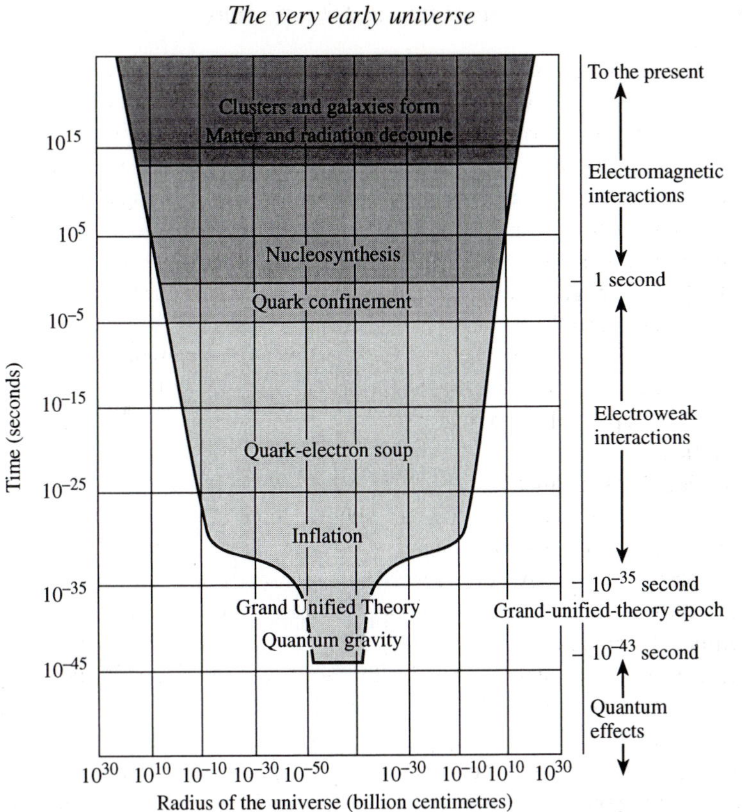

Fig. 5.6 Schematic picture of the evolution of the universe. In the 10^{-43} seconds following the big bang, quantum effects dominate and the four fundamental forces (electromagnetism, weak and strong nuclear forces, and gravity) are believed to have been unified in a single force. First gravity separates out, leaving the other three forces as a 'grand unified force'. When the strong nuclear force separates from the 'electroweak' force 10^{-35} seconds after the big bang, inflation begins. The matter in the universe consists of a 'soup' of quarks, which are the building blocks of protons and neutrons, and leptons, but the dominant form of energy is radiation. When the universe is one microsecond old, the quarks bind together to make protons and neutrons, and the weak nuclear and electromagnetic forces separate. Nucleosynthesis begins and continues until the universe is about three minutes old. When the universe is 300 000 years old, the matter cools sufficiently to become transparent to radiation and galaxies and clusters of galaxies can begin to form.

universe deviates from flatness, would have been smaller than 10^{-15} and at $t = 10^{-35}$ s it would have had to be smaller than 10^{-50}. The universe therefore had to be incredibly close to being flat at early times to have evolved to the kind of universe we see today. The inflationary phase leaves the universe in the required very flat state and in fact predicts that the average density of the universe today would be within one part in 10^4 of the critical value (eqn 4.24).

Several different versions of how inflation occurred have now been proposed. The essential common feature is the period of exponential expansion in the very early universe, which solves the horizon and flatness problems. However there is no evidence that any such phase ever occurred and it is in fact quite hard to see how such evidence could be obtained.

Extrapolating back still further in time, at $t \sim 10^{-43}$ s we reach what is known as the Planck time, when a quantum theory of gravity is required to extrapolate back any further in time. Although much effort has gone into producing a quantized theory of gravity, and thereby unifying all the forces of physics, there is no generally accepted theory of this type yet. The most successful of the attempts to date is the 'superstring' theory of Green and Schwartz. It seems unlikely that such theories will ever be testable.

Our current picture of the evolution of the universe is illustrated in Fig. 5.6.

5.5 Isotropy of the microwave background

The high degree of isotropy of the microwave background radiation both on the large and small angular scales has profound consequences for our picture of the universe. Let us consider the various kinds of anisotropy we might have expected and summarize the present observational situation for each.

360°, 24-hour, or 'dipole' anisotropy

The combined effects of the Earth's motion round the Sun, the Sun's motion round the Galaxy, the Galaxy's motion within the Local Group, the Local Group's motion with respect to the Virgo cluster, and the Virgo cluster's random motion with respect to the substratum should result in the Earth having a net motion with respect to the fundamental cosmological frame of reference or substratum. Even if the microwave background radiation were perfectly isotropic with respect to the cosmological frame, it should show a characteristic dipole anisotropy due to the effect of the Doppler shift on the intensity observed at the Earth. For radiation with a blackbody spectrum characterized by a temperature T, the observed intensity would still have a blackbody spectrum but with a temperature $T(\theta)$ depending on direction according to

$$T(\theta) = T_0(1 + \upsilon \cos\theta/c)/(1 - \upsilon^2/c^2)^{1/2} \tag{5.18}$$

where θ is the angle between the direction of observation and that of the Earth's motion with respect to the substratum and υ is the Earth's speed. The radiation appears hotter in the direction the Earth is travelling towards and colder in the opposite direction. Observations made with a telescope fixed on the Earth's surface will show a periodicity every 24 hours of sidereal time, corresponding to an angular scale of 360°.

This type of anisotropy was detected in 1977 by a group from the University of California, Berkeley, using a U2 aircraft, and confirmed by groups from Princeton and Florence using high-altitude balloons. When corrected for the

Earth's motion round the Galaxy and our Galaxy's motion through the Local Group, the velocity of the Local Group with respect to the microwave background radiation measured by COBE 622 ± 22 km s^{-1} in a direction given by Galactic coordinates $l = 277^\circ \pm 2^\circ$, $b = 30^\circ \pm 2^\circ$.

This seemed at first to be a disturbingly large speed for our random motion through space. However, studies based on redshift surveys of IRAS galaxies have shown that this motion can be understood as a result of the combined attraction of large clusters of galaxies within 100 Mpc of us, provided that the total value of Ω_0 is close to 1. Comparison with eqn (5.17) implies that over 95 per cent of the matter in the universe would then have to be non-baryonic.

180°, 12-hour, or 'quadrapole' anisotropy

If the universe were rotating or shearing (i.e. expanding anisotropically) then we might expect a large-scale anisotropy in which the background looks hotter or colder than average in two opposite directions on the sky. No such effect is observed and, apart from the dipole anisotropy mentioned above, the microwave background is isotropic on the large scale to an accuracy of 0.001 per cent.

Small-scale anisotropies

Since the universe is clearly inhomogeneous on mass-scales corresponding to galaxies and clusters of galaxies, say 10^8–10^{15} $M_\odot$, we expect that, at the epoch of recombination, density perturbations were present which could then grow under the influence of their own gravitation. For galaxies to have formed by the present epoch in a universe composed of normal, baryonic matter, we need the amplitude of these protogalactic density perturbations to be $\Delta\rho/\rho > 0.1$ per cent if the cosmological density parameter $\Omega_0 = 1$, and $\Delta\rho/\rho > 1$ per cent if $\Omega_0 = 0.1$. The corresponding temperature fluctuations in the observed background radiation would be $\Delta T/T \sim 10^{-4}$ on angular scales from $0.5'$ to $20'$. By 1980 observational limits gave $\Delta T/T < 5 \times 10^{-5}$ on angular scales from $3'$ to 1°, inconsistent with the condition for galaxy formation in a purely baryonic universe. Interest therefore focused on models in which most of the matter in the universe is in some dark, non-baryonic form, which would have decoupled from the radiation at a much earlier epoch than the baryonic matter. Non-baryonic matter can be broadly divided into two types: *cold dark matter* in which the particles are slowly moving at the epoch when they decouple from the radiation, and *hot dark matter* in which the particles are moving at speeds close to the speed of light when they decouple from the radiation. An example of hot dark matter would be a neutrino with non-zero rest mass. No examples of cold dark matter are known, but particle physicists have postulated the existence of particles which would be candidates, like the axion and the neutralino.

In 1992, the Cosmic Background Explorer (COBE) team announced that they had detected small-scale anisotropies in the microwave background radiation on scales of 10°. The amplitude of these fluctuations is about 1×10^{-5} and the scale of the fluctuations corresponded to structures 1000 Mpc in size today, far larger than any structure we have been able to study to date at the present epoch. Subsequent ground-based and balloon-borne experiments have confirmed the COBE result and extended it to smaller scales. The next generation of space-borne experiments should be able to detect these fluctuations on the scale of galaxies and clusters.

The detection of these fluctuations is a milestone for cosmology, because we can now test different scenarios for the nature of the dark matter and for the formation of galaxies. The fluctuations may well have their origin at the inflationary epoch, only 10^{-35} sec after the big bang.

The Sunyaev–Zeldovich effect

The hot gas in rich clusters of galaxies interacts with the photons of the cosmic microwave background by Compton scattering, with the result that towards a cluster the background looks cooler than average at long wavelengths and hotter than average at short wavelengths. The magnitude of this effect is about 0.001 K and has been measured towards a number of clusters.

5.6 The evolution of density fluctuations during the fireball

We can discuss a density fluctuation as a coherent entity only from the moment when it lies entirely within the horizon (Section 4.9). Prior to that the different parts of the fluctuation could not be in communication with each other. If $M_{\rm h}(t)$ is the mass within the horizon at time t, then

$$M_{\rm h}(t) \propto \rho(ct)^3 \propto t$$

during the radiation-dominated era since $\rho \propto R^{-4}$, and $R \propto t^{1/2}$ (Section 5.1). At the transitional epoch from the radiation- to matter-dominated eras, $t_{\rm eq}$, $M_{\rm h}(t_{\rm eq}) \sim 10^{15}(\Omega_0 h^2)^{-2} M_\odot$, where $h = H_0/100$, so fluctuations on the mass-scales of galaxies and clusters come within the horizon during the radiation-dominated era. Baryonic fluctuations then remain frozen at approximately their initial amplitude until recombination because the radiation and matter are locked together by Thomson drag. Non-baryonic matter fluctuations, on the other hand, grow steadily throughout this phase, with $(\Delta\rho/\rho) \propto t^{2/3}$ for all scales.

There are four types of density fluctuation: (a) *adiabatic*, which behave like sound waves, and in which the photon and matter fluctuations vary together; (b) *isothermal*, in which the matter density is perturbed but the temperature is not (i.e. we have matter fluctuations in a uniform photon bath); (c) *isentropic*, in which matter and radiation vary with opposite phases so that the total energy density remains constant; (d) *turbulent*, which are large-scale eddies in the coupled matter and radiation.

It was found by Silk that adiabatic fluctuations with

$$M < 2 \times 10^{12}(\Omega_0/\Omega_b)^{3/2}(\Omega_0 h^2)^{-5/4} M_\odot \tag{5.20}$$

where $h = (H_0/100)$, are damped out during the radiation-dominated era by photon diffusion out of overdense regions into underdense ones. Turbulent fluctuations are also damped out on lower mass-scales. In this case the energy associated with the damped out perturbations may have the effect of distorting the radiation from Planckian form (see next section).

Zeldovich and Harrison independently postulated that the amplitude of density perturbations at the moment they come within the horizon is $\Delta\rho/\rho \approx 10^{-4}$, independent of mass-scale (the *scale-free* hypothesis). It can be shown that the resulting matter density fluctuations at the epoch of recombination, on large scales, is then of the form

$$\Delta\rho/\rho = 10^{-4}\{M/M_{\mathrm{h}}(t_{\mathrm{eq}})^{-2/3}\}. \tag{5.21}$$

The Harrison–Zeldovich density fluctuation spectrum has been very successful in accounting for the observed hierarchy of galaxy clustering. The detailed evolution of the spectrum of density fluctuations depends on the cosmological parameters and on the nature of the non-baryonic matter in the universe. One of the successes of the hypothesis of an inflationary phase at $t \approx 10^{-35}$ s in the very early universe is that such models can generate fluctuations of the Harrison–Zeldovich form.

5.7 Distortions of the microwave background spectrum

If energy is injected into the radiation field at early epochs, for example by the dissipation of adiabatic or turbulent density fluctuations, the spectrum of the radiation may be expected to undergo distortion from a blackbody form. The mean temperature of the matter will be higher than that of the radiation and Compton scattering will then boost the photons to higher energies (and frequencies). Bremsstrahlung radiation from the hot gas will partly fill in the missing low-energy photons, and if the heat injection occurs at early enough epochs may even succeed in restoring a blackbody spectrum at the new, higher temperature of the matter.

Compton distortion of the background spectrum may also occur after the epoch of recombination if, for example, as the result of the formation of galaxies or quasars, the intergalactic gas (assuming some exists) is reionized.

The observed spectrum of the microwave background measured by the COBE satellite (Fig. 5.1) is Planckian to an accuracy of 0.005 per cent over the wavelength range 0.05–5 cm, so any distortion must be extremely small. This also rules out the existence of any significant population of pre-galactic objects radiating at submillimetre wavelengths. Any possibility that the microwave background is due to radiation from relatively local dust grains is also eliminated.

5.8 Problems

5.1 Show that the substitution $\nu' = \nu/Z$ into the expression for the intensity of blackbody radiation $I(\nu)$ given by eqn (1.2) results in a blackbody spectrum corresponding to a temperature $T' = T/Z$. Give an interpretation of this result.

5.2 Let t_1 be the epoch of electron–positron annihilation when $T = 5 \times 10^9$ K, and let t_2 be the epoch when $T = 10^9$ K and fusion of helium begins. Calculate t_1 and t_2 for the two cases:

(a) the universe is radiation-dominated for $t < 3 \times 10^5$ yrs, then matter-dominated until the present epoch ($t_0 = 10^{10}$ yrs);

(b) the universe is radiation-dominated for $t < t_1$, then matter-dominated till the present epoch.

If $X_n = X_p = 0.5$ at $t = t_1$, calculate the neutron abundance at t_2 assuming neutron decay according to eqn (5.16) for a time $(t_2 - t_1)$, for the two cases (a), (b). Hence deduce the helium abundance in each case, assuming all surviving neutrons are used to make helium-4 nuclei (2 neutrons + 2 protons).

Comment on the significance of these results for the observed universe.

6
From the fireball to the present

6.1 Before galaxies formed

The epoch of recombination at $z_{\rm rec} \approx 1000$ marks the moment when baryonic matter is at last able to evolve independently of the radiation. A region where the matter has a slightly higher density than average will have its expansion slowed down and it will start to separate out from the surrounding matter. If the initial density excess $\Delta\rho/\rho$ is large enough, the expansion of the fragment will eventually be halted by its self-gravity and it will collapse together to form a galaxy or cluster of galaxies depending on the mass of the fragment. The time for this expansion and collapse is

$$t_c \approx 10^6 h^{-1} \Omega_0^{-1/2} (z_{\rm rec}/1000)^{-3/2} (\Delta\rho/\rho)^{-3/2} \text{ years} \tag{6.1}$$

so the strongest perturbations form first.

If, as is currently believed to be the case, non-baryonic matter perturbations are present and have already evolved significantly by the epoch of recombination, then at recombination the baryons will start to fall into the potential wells defined by the non-baryonic perturbations. The baryonic density fluctuations quickly catch up the non-baryonic fluctuations in amplitude and the two types of matter perturbation then evolve together (Fig. 6.1).

The detailed evolution of the density fluctuation spectrum depends on the cosmological density parameters and on the nature of the dark matter. Figure 6.2 shows the form of the density fluctuation spectrum on different scales at late times for three different scenarios, each of which starts initially from a spectrum of the Harrison–Zeldovich form. On large scales, the spectrum has the primordial form $\Delta\rho/\rho \propto M^{-2/3}$ in each case. Free-streaming of hot dark matter particles has completely suppressed structure on small scales in this case.

Density fluctuations present at the epoch of recombination can form into galaxies provided they satisfy an important criterion discovered by Jeans. The time-scale for gravitational contraction must be less than the time-scale for sound waves to cross the irregularity; otherwise pressure forces will prevent the collapse. This means that the size of the fluctuation must be greater than the distance travelled by a sound wave in one free-fall time ($\sim (G\rho)^{-1/2}$)

$$L \sim \upsilon_{\rm s} (G\rho)^{-1/2} = L_{\rm J}, \tag{6.2}$$

where $L_{\rm J}$, is the Jeans length, $\upsilon_{\rm s}$ is the velocity of sound, and ρ is the density of

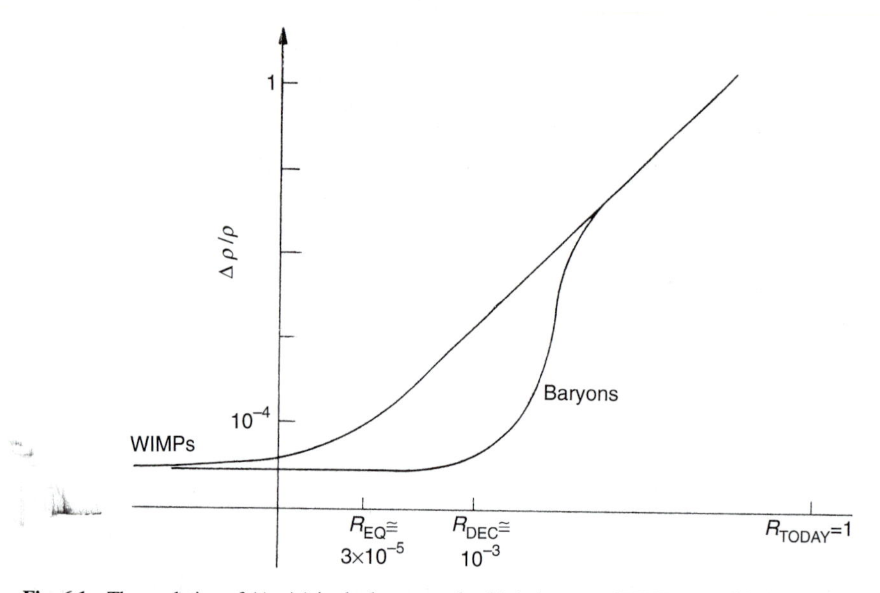

Fig. 6.1 The evolution of $(\Delta\rho/\rho)$ in the baryon and cold dark matter (WIMP or weakly interacting massive particle) components. The perturbations in the WIMPs begin to grow at the epoch of matter–radiation equality. However, the perturbations in the baryons cannot begin to grow until just after decoupling. After decoupling baryons fall into the WIMP potential walls, and within a few expansion times the baryon perturbations 'catch up' with the WIMP perturbations.

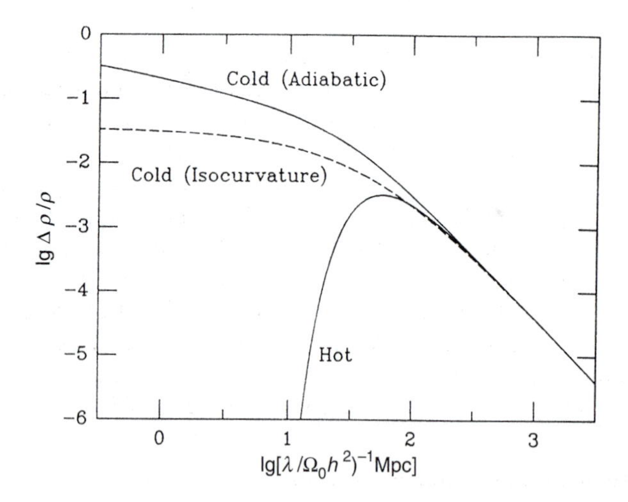

Fig. 6.2 The density fluctuation spectrum on different scales, $\Delta\rho/\rho(\lambda)$, specified at some time after the epoch of matter–radiation equality, for hot and cold dark matter with constant-curvature, adiabatic fluctuations, and for cold dark matter with isocurvature, Harrison–Zeldovich fluctuations. For all three cases $\Delta\rho/\rho \propto \lambda^{-2} \propto M^{-\frac{2}{3}}$ for $\lambda \gg \lambda_{EQ}$. The overall normalization is arbitrary.

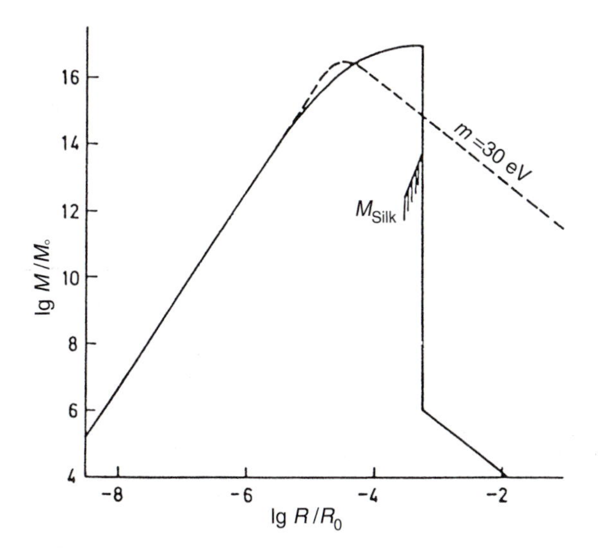

Fig. 6.3 Plot of the Jeans mass versus expansion factor. The solid line is for a photon and baryon fluid of ideal gases. The dotted line is for collisionless (hot dark matter) particles of $m = 30$ eV.

the gas. The corresponding critical (or *Jeans*) mass at the epoch of recombination is (Fig. 6.3)

$$M_J \sim 10^6(\Omega_0 h^2)^{-1/2} M_\odot \tag{6.3}$$

Fluctuations on a mass scale slightly larger than this, if any are present, will complete their collapse in a few million years. They might collapse directly to black holes or might fragment into stars. One proposal is that this might be the origin of globular clusters. Another is that this might give birth to a transient generation of massive stars which produce radiation and heavy elements. This is one way of explaining the metals present in the oldest stars in our Galaxy, which never seem to be found with a heavy element abundance $\lesssim 10^{-4}$ relative to hydrogen. If there was a pregalactic generation of radiating objects, whether stars or accreting black holes, then they might have the effect of reionizing the remaining gas. The failure to detect any distortions to the Planckian spectrum of the microwave background by COBE does, however, strongly limit the amount of radiation that can have been emitted by pregalactic objects.

6.2 Formation of galaxies and clusters

There are several pictures of galaxy and cluster formation which are consistent with the high degree of isotropy of the microwave background. We saw in Section 5.5 that this isotropy eliminates the possibility that galaxies and clusters

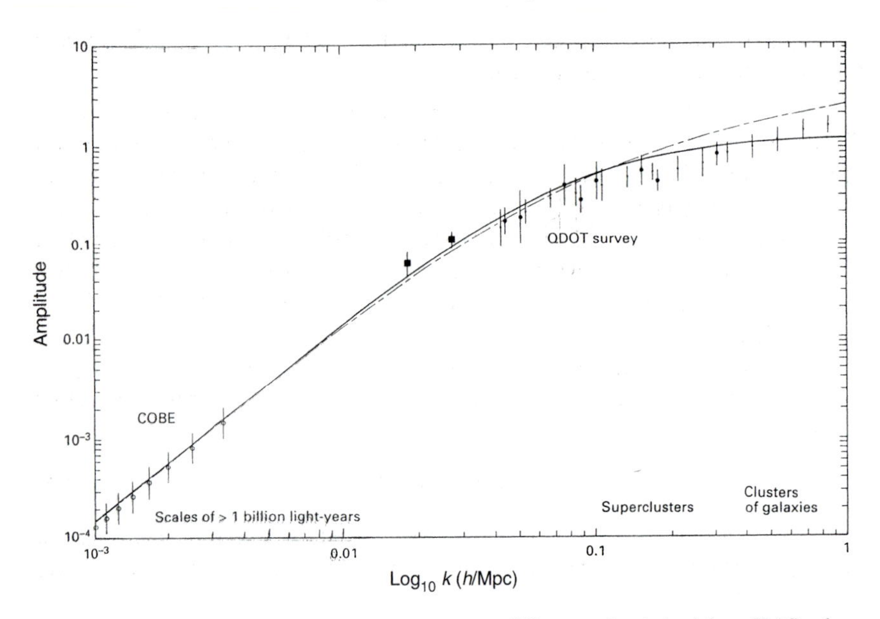

Fig. 6.4 The observed density fluctuation spectrum on different scales derived from IRAS galaxy redshift surveys and from the COBE microwave background fluctuations. The theoretical models are for a model with both hot and cold dark matter (mixed dark matter, solid curve) and for a simple cold dark matter model.

were formed by gravitational means in a purely baryonic universe. The main current theories are:

(a) *adiabatic fluctuations in a cold dark matter universe*: the cold dark matter condenses in lumps on small scales which aggregate together to form galaxy halos, in the cores of which the baryonic matter settles to make the visible parts of galaxies. Galaxies then cluster together under the influence of gravity to form clusters and superclusters ('bottom-up' scenario).

(b) *adiabatic fluctuations in a hot dark matter universe*: the largest mass structures (superclusters) form first as 'pancakes' and then fragment to give galaxies ('top-down' scenario). Adiabatic fluctuations in a baryonic universe follow a similar evolution, but would require too large fluctuations in the microwave background to be consistent with observations.

(c) *isothermal fluctuations*, in which matter fluctuations are superimposed on a uniform radiation field are now considered unphysical, but *isentropic fluctuations*, in which the total energy density remains uniform, have been advocated by Peebles as a possible baryonic universe model. These result in a bottom-up scenario. Other *ad hoc* models include the *cosmic explosion scenario*, in which the observed voids are due to astrophysically generated explosions, e.g. in quasars, and galaxies form in the resulting shock fronts; and *cosmic strings*, in which galaxy formation is aided by linear topological defects left over from phase transitions in the very early universe.

The density fluctuation spectrum observed on large scales allows us to distinguish between these different scenarios. It seems that no simple scenario can fit both the COBE microwave background fluctuations and the statistics of large-scale galaxy clustering (see below). Scenarios that are being pursued at the moment include a model with both hot and cold dark matter (*mixed* dark matter, see Fig. 6.4), a low-Ω cold dark matter model with non-zero cosmological constant (see Chapter 8), and cold dark matter models in which the initial perturbation spectrum is flatter than the Harrison–Zeldovich form.

The clustering of galaxies can be described by the *covariance function*, $\zeta(r)$, defined as the excess probability of finding a galaxy at a distance r from a random galaxy:

$$\mathrm{d}p = n[1 + \zeta(r)]\mathrm{d}V \tag{6.4}$$

where n is the average number-density of galaxies and $\mathrm{d}V$ is a small element of volume. Peebles and coworkers found from studies of a variety of galaxy catalogues that

$$\zeta(r) \sim (r/10)^{-1.8} \tag{6.5}$$

where r is in Mpc.

Numerical simulations of galaxy clustering show that eqn (6.5) is indeed roughly what would be expected if the fluctuation spectrum at recombination were of the form (5.21).

Once protogalactic fragments start to separate out from the surrounding gas, their ultimate fate depends on how efficiently star formation proceeds during the collapse phase. If all the gas in the fragment has formed into stars before the fragment collapses together, then an elliptical galaxy is formed. Otherwise a disc of gas supported by centifugal force and surrounded by an extended ellipsoidal component is formed, i.e. a spiral galaxy. Models have been constructed which account in detail for the luminosity and colour distributions, the rotation curves, and metal abundance gradients seen in the different galaxy types.

It is now believed that interactions and mergers between galaxies play a major role in the evolution of galaxies. Many elliptical and lenticular galaxies (perhaps all) may have been the result of mergers of gas-rich systems.

6.3 Intergalactic gas in clusters

The discovery that rich clusters of galaxies are powerful X-ray emitters due to hot ($\sim 10^8$ K) intergalactic gas has important implications for galaxy evolution. The detection of X-ray line emission due to ionized iron in these cluster sources demonstrates not only the presence of the hot, Bremsstrahlung-emitting gas, but also shows that much of this gas has been processed in stellar interiors. Three mechanisms have been proposed for getting this processed gas out of the galaxies. The first uses supernova-driven winds to blow the gas out of the galaxies. The second is based on tidal interaction between two galaxies during a close encounter. And the third involves the stripping of gas from the galaxies as they travel through the intergalactic gas under the action of the cluster gravitational field. The latter mechanism requires either a significant amount of intergalactic gas in the cluster initially, or one of the other mechanisms to have been operating first.

Stripping of gas from spiral galaxies in clusters is a promising explanation of the origin of lenticular (S0) galaxies, but there is still debate about whether all lenticulars are gas-denuded spirals in origin (probably not). Clusters of galaxies at large redshift do seem to show a higher proportion of blue, presumably spiral, galaxies than similar clusters at the present epoch. Most rich clusters at the present epoch are composed predominantly of ellipticals and lenticulars.

6.4 The masses of galaxies and clusters

We saw in Chapter 4 that whether the universe keeps on expanding indefinitely, or ultimately falls back together into a second fireball, depends on whether the average density of matter is less than or greater than the critical value

$$\rho_{ES} \sim 5 \times 10^{-27} (H_0/50)^2 \text{ kg m}^{-3} \tag{6.6}$$

(equivalent to $\Omega_0 = 1$, the Einstein–de Sitter value—Section 4.7). We now try to determine the average density of matter in galaxies and in other possible forms.

To find the average density of matter in galaxies we first need to determine the average mass of a galaxy and then multiply by the average number of galaxies per unit volume, determined by galaxy counts (Section 7.8).

A variety of methods have to be used to determine galaxy masses, depending on the galaxy type.

(a) Spirals

Most of the material in the disc of spirals is moving in an approximately circular orbit in a balance between centrifugal force and gravity. Hence for material far out from the centre of the galaxy

$$V^2/r \sim GM/r^2 \quad \text{or} \quad M \sim rV^2/G \tag{6.7}$$

The rotation curve $V(r)$ can be determined by observing the Doppler-shifted 21-cm line of neutral hydrogen (Fig. 6.5). However, we see that for M31 the rotation curve does not drop down in the outer parts of the galaxy as predicted by eqn (6.7), pointing to the existence of a substantial halo of non-luminous material surrounding the visible galaxy, containing 90 per cent of the total mass of the galaxy. A similar argument applies to our own Galaxy. Theoretical arguments show that a disc galaxy would be unstable to the formation of a bar unless a

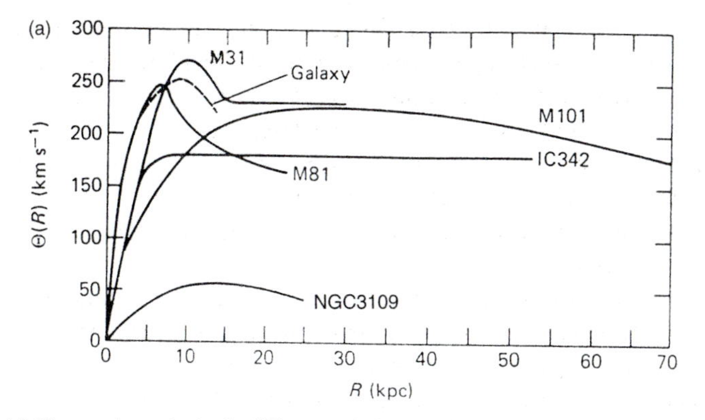

Fig. 6.5 (a) The rotation velocity in different galaxies as a function of distance from the centre.

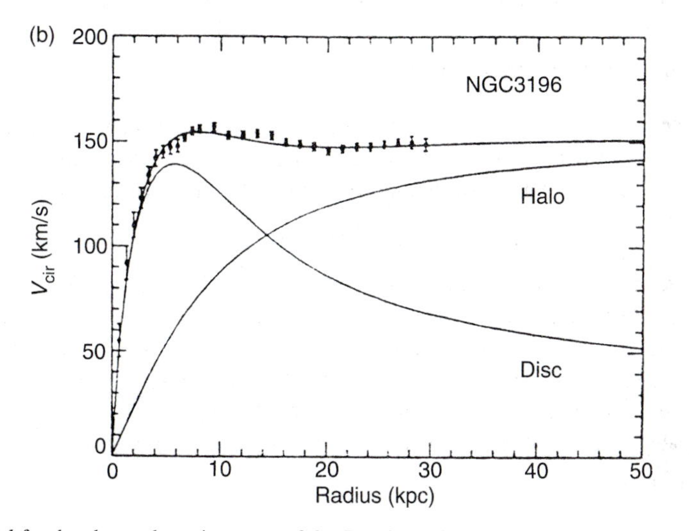

(b) Model for the observed rotation curve of the Sc spiral galaxy NGC3198, showing contributions from the visible disc of stars and gas and from the dark halo. Figure from van Albada *et al.* (1985), *Astrophysics Journal* **295**, p. 305.

massive halo surrounds the disc. It is not known what form the mass in this halo takes, but it could be dwarf stars of very low luminosity (brown dwarfs), black holes, or exotic particles like *axions* or *neutralinos*.

(b) Ellipticals

Here we have to use what is known as the *virial theorem*, which tells us that for a system in equilibrium $2 \times$ kinetic energy $=$ gravitational energy, or $MV^2 \sim GM^2/r$ which is equivalent to eqn (6.1). We estimate the average kinetic energy per unit mass from the Doppler spreading of the emission lines in the spectrum of the nucleus of the galaxy. Again, we find that ellipticals are surrounded by a halo of dark matter, containing 90 per cent of the total mass of the galaxy.

(c) Mass-to-light ratio

By observing the ratio of mass to light for galaxies of different types, e.g. in binary systems, we can estimate the masses of individual galaxies from their total luminosity. If this ratio is determined over the optically visible extent of a galaxy, a typical value of $M/L \sim 3-10$, in solar units, is found for spirals and 10–80 for ellipticals. However the virial theorem applied to binary galaxies, and to groups of galaxies, yields $M/L \sim 100$, supporting the concept that spiral galaxies are surrounded by massive halos of low-luminosity material, in the form of dwarf stars, black holes, or any other material that does not contribute significantly to the light of the galaxy.

Methods (b) and (c) can be used to estimate the masses of clusters of galaxies. The virial theorem becomes

$$M = \bar{r}\overline{V^2}/G, \tag{6.8}$$

where the averages indicated by bars are over all the galaxies in the cluster, and r now refers to the distance between the galaxies. Typical masses are in the range 10^{13}–$10^{15}\ M_\odot$ and typical mean densities inside clusters are 10^{-25} kg m^{-3}.

Discrepancies between methods (b) and (c), with virial theorem masses up to ten times greater than visible masses ($M/L \sim 100-500$), again suggest 'missing matter'. Both extended X-ray emission from rich clusters and the appearance of radio 'tails' within clusters that there is intergalactic gas in clusters (Section 2.11). Recent maps of this gas with the ROSAT satellite have suggested that the mass of this gas is significantly greater than the mass in the cluster galaxies and may constitute as much as 30 per cent of the dynamical estimates of the total mass of the cluster. If clusters are representative of the average distribution of matter in the universe, this would imply that Ω_0 must be $\ll 1$, since we know the contribution of baryons to the density of the universe is only $0.05\,(50/H_0)^2$. For $\Omega_0 = 1$, we require either that the dynamical estimates of total cluster mass are underestimated, or that the X-ray gas masses are overestimates, or that baryons are somehow funnelled to the cores of rich clusters.

6.5 The average density of matter in the universe due to galaxies

The luminosity function of galaxies, i.e. the number of galaxies per unit volume having luminosities in the range $(L, L + \mathrm{d}L)$, can be well represented by

$$\phi(L)\,\mathrm{d}L = \phi_*(L/L_*)^{-\alpha}\exp(-L/L_*)\,\mathrm{d}L \tag{6.9}$$

where ϕ_* is a constant, $L_* = 3.2 \times 10^{10}\,L_\odot$, and $\alpha \cong 1.25$. Estimates of the total

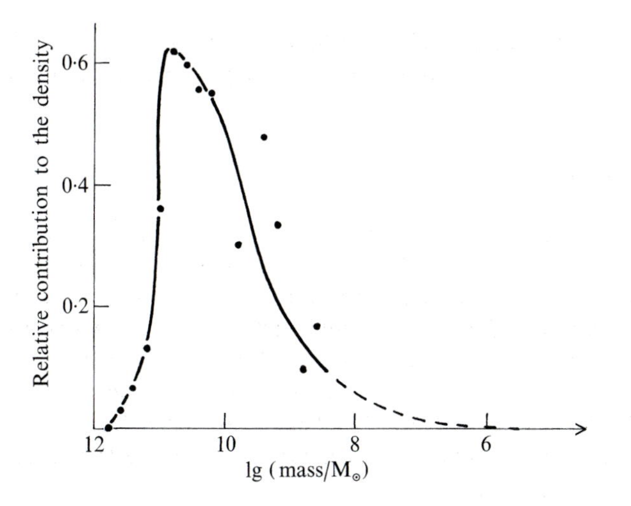

Fig. 6.6 The relative contribution of galaxies of different masses to the average density of matter in the universe.

luminosity density of galaxies in the blue band lie in the range $5-15 \times 10^7 L_\odot$ Mpc^{-3} (assuming $H_0 = 50$). Using $M/L = 100$, we find

$$\rho_{g,0} \sim 5 \times 10^{-28}(H_0/50)^2 \text{ kg m}^{-3} \qquad (6.10)$$

or $\Omega_{g,0} \sim 0.1$, with an uncertainty of at least a factor of 2 either way. This estimate is a factor of 10 below the critical value (eqn 6.6). It is however comparable with the density of baryonic matter derived from primordial nucleosynthesis arguments (eqn 5.17).

Fig. 6.6 shows the relative contribution to the average density of the universe of galaxies of different masses. Most of the mass comes in the form of high-mass galaxies, 10^{10}–10^{12} $M_\odot$. The curve has been extrapolated right down to isolated globular clusters, although it is not certain that the curve does not peak up again at this level.

An important development of the past decade has been the growth of large-area galaxy redshift surveys, which allow large-scale dynamical estimates of Ω_0. Redshift surveys based on the IRAS 60 μm survey have been especially influential, because for the first time they give galaxy samples covering the whole sky and unaffected by extinction by dust. These surveys have been used to map the galaxy density distribution and hence estimate Ω_0 both by calculating the net attraction of the galaxies and clusters within 150 h^{-1} Mpc and comparing this to the peculiar velocity of our Galaxy, and by comparing the peculiar velocities of galaxies in the field with that predicted from the galaxy density distribution. From both types of study, values of Ω_0 close to the critical value of 1 have been generally found.

6.6 Some other possible forms of matter

The best candidate for matter with a significant fraction of the critical density (eqn (6.6)) is intergalactic gas, since we do not expect galaxy formation to be a 100 per cent efficient process. This will be discussed in the next section. Some other possible forms are discussed below.

Dwarf galaxies

Although Fig. 6.6 suggests that the contribution from dwarf galaxies to the average density of matter in the universe is small, very few of these objects are known (in fact all are in the Local Group) so their number density is very uncertain. The estimate of Fig. 6.6 would have to be out by a factor of more than 100 to arrive at the critical density. Whether intergalactic globular clusters and isolated stars exist is still unknown. A direct measurement of the optical background radiation might help to decide this.

Compact objects and quasars

The simplest view about quasars and other compact objects (e.g. N-galaxies) is that they represent outbursts in the nuclei of galaxies. As such, their contribution to the average density of matter would already have been included in eqn (6.10).

Even if they are a distinct class of object from galaxies, quasars make a negligible contribution to the average density of matter, unless a significant proportion of them are 'local' objects with intrinsic redshifts.

Dead galaxies and black holes

The normal types of galaxy surveyed in Section 2.6 can be expected to retain their present appearance for a time much longer than the Hubble time. However, there is the possibility of an earlier generation of galaxies now no longer visible due to exhaustion of stellar nuclear energy sources (they would have to have formed only massive stars). These dead galaxies would now be in the form of a cluster of neutron stars or black holes.

A related possibility is that strong density fluctuations present in the very early universe led to the formation of primordial black holes. Hawking has introduced the concept of very low-mass black holes, of mass down to 10^{-8} kg, and shown that these would rapidly evaporate and explode through a quantum gravitational effect close to the black hole's event horizon (the boundary between events that we can in principle observe and those that we can never observe). A black hole of mass M evaporates through the radiation of photons, elementary particles, and gravitons, on a time-scale

$$t(M) \sim 10^{10}\,(M/10^{12}\text{ kg})^3 \text{ years.} \qquad (6.11)$$

The effect of Hawking radiation is therefore negligible for black holes of solar and galactic mass. A black hole of 10^{12} kg should be exploding now in a burst of gamma rays and other particles. Our failure to detect such events and limits from the observed gamma-ray background put strong bounds on the contribution of such objects to the density of matter in the universe at the present epoch ($\Omega < 10^{-8}$). However, black holes in the mass range 10^{13} kg–$10^7 \, M_\odot$ could contribute a small fraction ($\sim$ 1 per cent) of the critical density.

Planets, rocks, and dust

Only weak limits can be placed on the amount of cool, solid matter in the universe. If the average density of the grains is $\rho_d \sim 10^3$ kg m^{-3}, and their contribution to the average density of the universe is given by eqn (6.6), then in order to be able to see a Hubble distance without drastic absorption, the grains need to be larger than 1 mm:

$$\frac{4\pi}{3} a^3 \rho_d \eta_d \sim 5 \times 10^{-27}, \quad \eta_d \pi a^2 c \tau_0 < 1, \tag{6.12}$$

where a, η_d are the radius and number density of the grains. Thus there is no real observational limit on the amount of matter in the universe in the form of large dust, rocks, or planets.

Neutrinos

We mentioned briefly in Section 5.3 that cosmic neutrinos and anti-neutrinos are expected to become decoupled from matter and radiation at an early stage in the fireball. The expected energy density is about 20 per cent of that of photons, i.e. negligible compared to matter at the present epoch. This neutrino flux is completely undetectable with present techniques (Section 1.3). Neutrinos could make a dominant contribution to the matter in the universe if they have a non-zero mass of $\gtrsim 10$ eV/c^2.

Other non-baryonic matter

Theories which seek to unify the forces of physics during the very early universe have no difficulty in generating new particles, which may be capable of supplying a significant fraction of the matter in the universe. Of especial interest are particles which are moving slowly at the time they decouple from the radiation ('cold dark matter'), for example axions or neutralinos. Searches are under way in the US and Europe to try to detect such particles.

Gravitational radiation

It would be possible for this to be the dominant form of energy at the present epoch: current gravitational wave detectors are too insensitive by a large factor to test this. However, the isotropy of the microwave background implies that gravitational radiation would have to have wavelengths smaller than 8 Mpc if it contributed the critical density.

Cosmic rays

The local energy-density of cosmic rays ρ_{cr} is about the same as that of the microwave background and corresponds to

$$\begin{aligned} \rho_{cr} &\sim 1.6 \times 10^{-30} \text{ kg m}^{-3} \\ &\ll \rho_{ES} \end{aligned} \tag{6.13}$$

Even this value may hold only within our Galaxy. In fact cosmic-ray nuclei with energies $< 10^{17}$ eV are confined to our Galaxy by its magnetic field, so are almost certainly of Galactic origin. And the average energy density of cosmic-ray electrons must be less than a thousandth of the local value, otherwise inverse Compton interaction with the microwave background and integrated starlight photons would give too large an X-ray background (Section 1.4). However, the higher-energy cosmic-ray nuclei leak out of our Galaxy in only about 3×10^6 years (this can be deduced from the abundance of the cosmic-ray-created elements Li, Be, B), so the density (eqn (6.13)) may well be universal. If it is, then we cannot have intergalactic gas at the critical density (eqn (6.6)), since (i) the cosmic rays would heat it up so much that too many X-rays would be produced, and (ii) the production of X-rays from cosmic ray secondary π^0 meson decay would exceed the observed background in the range 50–100 MeV.

6.7 Intergalactic (and intercluster) gas

If there is a uniform intergalactic (and intercluster) gas with a density comparable to the critical density (eqn (6.6)), then rather stringent limits can be set on its physical state.

Atomic hydrogen (H_I)

This would be observable by the following.

(a) 21-*cm absorption* : failure to see this in the radio galaxy Cyg A and in other radio galaxies shows that the density of intergalactic neutral hydrogen is

$$\rho_{H_I} < 10^{-28} \text{ kg m}^{-3}.$$

(b) 21-*cm emission:* again, this has not been observed.

(c) *Soft X-ray absorption:* an X-ray photon ionizes a hydrogen (or helium) atom, which then recombines, but emits a less energetic photon. The opacity of hydrogen for $\lambda < 912$ Å is proportional to $(\lambda/912)^3$, and failure to see absorption down to 50 Å shows that

$$\rho_{H_I} < 10^{-28} \text{ kg m}^{-3}.$$

(d) *Lyman-α absorption:* a trough should appear on the short wavelength side of the Lyman-α line at 1216 Å. For quasars this is conveniently red-shifted into the visible range, and the absence of such a trough shows that for uniformly distributed atomic hydrogen

$$\rho_{\mathrm{H_I}} < 10^{-33}\ \mathrm{kg\ m^{-3}}.$$

However a forest of absorption lines is often detected in the spectra of high-redshift quasars shortward of Lyman α and these are believed to be due to a pervasive distribution of low-density neutral hydrogen clouds, but with an average density well below the critical value ($\Omega \sim 0.003$).

Ionized hydrogen ($\mathrm{H_{II}}$)

We are likely to have $T > 10^5$ K since a cold, ionized gas recombines very quickly, and ionizing mechanisms, e.g. ultraviolet photons or cosmic rays produced by galaxies and quasars, tend to heat the gas too. The reheating of the gas cannot have occurred too soon after decoupling (Section 5.2), since observable distortions in the 2.7 K blackbody spectrum would have been produced.

In a hot ionized gas free electrons moving under the influence of each others' electrostatic fields radiate *free–free* (or *thermal Bremsstrahlung*) radiation. The observed radio and X-ray backgrounds give strong limits on the temperature:

Radio (20 cm): the gas can only have been heated to 10^5–10^6 K at redshifts $z < 100$.

Hard X-rays ($E > 1$ keV): $T \lesssim 3 \times 10^8$ K if $\rho = \rho_{\mathrm{ES}}$. The 1–100 keV background has been attributed to 4×10^8 K gas with $\rho \sim 0.5\rho_{\mathrm{ES}}$ (see Section 7.10).

Soft X-rays ($E = 0.25$ keV): there is a detection of gas with

$$\rho \sim \rho_{\mathrm{ES}}, \quad T \sim 10^6 K,$$

but this is due to a halo of hot gas round our own Galaxy. This hot intergalactic gas attenuates the light from distant sources (but not the integrated background) by Thomson (i.e. free-electron) scattering, by a factor $\exp(-\tau_{\mathrm{e}})$, where

$$\tau_{\mathrm{e}} = 0.023\{(1+z)^{3/2} - 1\}$$

in the Einstein–de Sitter model.

Other weak effects are the absorption of low-frequency radio waves, (Faraday) rotation of the plane of polarization of distant sources if there is also an intergalactic magnetic field, and a frequency-dependent time lag in the arrival of light from variable sources (*dispersion*).

Molecular hydrogen

Intergalactic molecular hydrogen would be expected to absorb ultraviolet radiation in the Lyman and Werner bands. The absence of an absorption trough on the short-wavelength side of Lyman α shows that molecular hydrogen cannot make up a very significant fraction of the matter in the universe, although it has recently been recognized that it makes up a major part of the gas in our Galaxy.

6.8 Problems

6.1. Test how well the galaxies of the Local Group (Table 1.1; p. 4) agree with Fig. 6.6. What reasons can you think of for less than perfect agreement?

6.2. Apply the virial theorem (eqn (6.8)) to the Local Group of galaxies (assume the total velocity of the galaxies is $\sqrt{3}$ times the radial velocity). How well does the virial-theorem mass agree with the total observed mass, and why?

7
Observational cosmology

7.1 Introduction

The isotropic 2.7 K blackbody radiation, the 24 per cent cosmic helium abundance, and the similarity between the ages of galaxies and the age of the universe all support the big-bang models derived from general relativity assuming the cosmological principle. Can we test these models in more detail? In particular, can we determine the current value of the deceleration parameter q (Section 4.7) and thereby deduce what the future of the universe will be? In this chapter we look at a variety of cosmological tests, all using discrete sources of radiation, that have been applied to try to answer this question. The answer turns out to be inextricably bound up with the *evolution* of the different classes of discrete source in the universe.

The main tests involve comparing luminosity and diameter distance (Section 3.2) with redshift, source counts (Sections 1.7 and 3.5), and integrated background radiation (Section 1.8). It was radio-source counts that first showed that strong evolution must affect some populations of source.

7.2 Newtonian theory

Suppose that we are in an expanding Newtonian universe, in a Euclidean geometry. The light from a source at distance d, receding with velocity $v = H_0 d$ (where H_0 is the Hubble constant—Section 3.3), will suffer a Doppler shift

$$\frac{\nu_e}{\nu_0} = \frac{\lambda_0}{\lambda_e} = 1 + \frac{\Delta\lambda}{\lambda_e} = 1 + z = Z, \tag{7.1}$$

where λ_e, ν_e, and λ_0, ν_0 are the wavelengths and frequencies of emission and observation, respectively, and $\Delta\lambda = \lambda_0 - \lambda_e$, of magnitude

$$z = v/c. \tag{7.2}$$

The flux S from the source is related to its luminosity P by the inverse-square law

$$S = P/d^2, \tag{7.3}$$

and the apparent angular size θ (rad) of an object of linear size l is given by

$$\theta = l/d. \tag{7.4}$$

The number of sources per steradian with distances $< d$ is $\eta d^3/3$, where η is the number of sources per unit volume (the *number density*), and if all the sources have the same luminosity P, then the number of sources per steradian with fluxes brighter than S is

$$N(S) = \frac{1}{3}\eta(P/S)^{3/2}. \tag{7.5}$$

The relations (7.2)–(7.5) might be expected to hold generally for $v \ll c$, i.e. $z \ll 1$, but we know they will break down as $v \to c$ because of the effects of special relativity.

7.3 Special relativity cosmology: the Milne model

We now take into account the effects of special relativity, but neglect the effects of gravitation. If we consider particles moving with the substratum, observing them from an inertial frame, then there are no forces acting on them so they all move with uniform velocity with respect to each other. A fundamental observer sitting on one of these particles continues to use Euclidean space coordinates, and measures the velocity of a particle at position vector $\boldsymbol{r}$ at time t to be $\boldsymbol{v}(\boldsymbol{r}, t)$. The only motion consistent with the cosmological principle turns out to be

$$\boldsymbol{v}(\boldsymbol{r}, t) = \boldsymbol{r}/t \tag{7.6}$$

(from eqn (4.8); this corresponds to taking $R(t) \propto t$, as we expect from Section 4.6). All particles would have been at the origin at $t = 0$ and then expand out isotropically with uniform velocity.

The Doppler shift can be shown to be

$$\frac{\nu_e}{\nu_0} = Z = \frac{1 + v/c}{(1 - v^2/c^2)^{\frac{1}{2}}} \tag{7.7}$$

(which agrees with the Newtonian expression (7.2) provided $v/c \ll 1$), and the flux from a distant source is now given by

$$(P/r^2)Z^{-4}, \tag{7.8}$$

where r is the distance of the source at the moment of emission, the factor Z^{-4} taking account of the various special relativistic effects of the source's motion on its apparent brightness. If the signal is received at time t_0, it was emitted at time $t_0 - r/c$, and eqn (7.6) implies that

$$r = v(t_0 - r/c), \quad \text{or} \quad r = vt_0/(1 + v/c).$$

Eqn (7.8) becomes

$$S = \frac{P}{(ct_0(z + z^2/2)^2}, \tag{7.9}$$

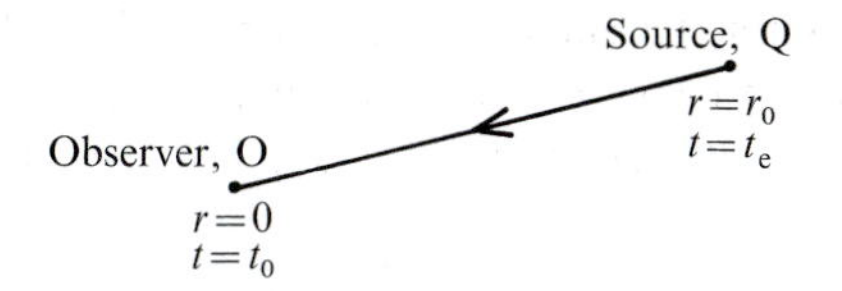

Fig. 7.1 The radial null geodesic linking the events of emission and reception of a signal sent from Q to O.

using eqn (7.7). The number of sources per steradian, with redshift factors less than or equal to Z is found to be

$$N(Z) = \eta_0(ct_0)^3(Z^2/8 - 1/8Z^2 - \frac{1}{2}\ln Z),$$

where η_0 is the local number density of sources at the present epoch.

Although the Milne model could be a reasonable approximation at the present epoch (we saw in Section 4.6 that all general relativity models with $k = -1$ tend to the Milne model for large t), it cannot be valid back to $t = 0$, since the density of matter becomes infinite as $t \to 0$ so that gravitational effects cannot be neglected.

7.4 General relativistic cosmology: the redshift

Our starting point is the Robertson–Walker metric (Section 4.5):

$$ds^2 = dt^2 - \frac{R^2(t)}{c^2}\left(\frac{dr^2}{1 - kr^2} + r^2 d\theta^2 + r^2 \sin^2\theta \, d\phi^2\right), \tag{7.11}$$

together with the fact that for two events (r, θ, ϕ, t), $(r + dr, \theta + d\theta, \phi + d\phi, t + dt)$ connected by a light signal, the interval

$$ds = 0 \qquad \text{(Section 4.4)}.$$

Consider a photon emitted by a source at Q at time t_e, so that the event of emission is $(r_0, \theta_0, \phi_0, t_e)$ and support that the photon is received by an observer at the origin at time t_0, so that the event of observation is $(0, \theta_0, \phi_0, t_0)$ (see Fig. 7.1).

Clearly the light signal travels in a radial straight line, by symmetry, so $d\theta = d\phi = 0$ and

$$ds^2 = 0 = dt^2 - \frac{R^2(t)dr^2}{c^2(1 - kr^2)}$$

for any element of the light ray joining Q to O, or

$$\frac{dr}{(1 - kr^2)^{1/2}} = -\frac{c dt}{R(t)} \tag{7.12}$$

for an incoming signal. Integrating this from Q to O:

$$\int_0^{r_0} \frac{\mathrm{d}r}{(1-kr^2)^{1/2}} = \int_{t_e}^{t_0} \frac{c\mathrm{d}t}{R(t)}. \tag{7.13}$$

Now recall that we chose r to be a *co-moving* coordinate (Section 4.5), so that at a later time $t_0 + \mathrm{d}t_0$ the source is still defined by $r = r_0$ (its change in distance is incorporated into the scale factor $R(t)$). The left-hand side of eqn (7.13) therefore does not change with time for a particular source–observer pair. Now consider a second signal emitted by Q at time $t_e + \mathrm{d}t_e$, and suppose this later signal is received by O at $t_0 + \mathrm{d}t_0$. Then eqn (7.13) becomes

$$\int_0^{r_0} \frac{\mathrm{d}r}{(1-kr^2)^{1/2}} = \int_{t_e+\mathrm{d}t_e}^{t_0+\mathrm{d}t_0} \frac{c\mathrm{d}t}{R(t)} \tag{7.14}$$

for this later signal. Subtracting eqn (7.14) from (7.13),

$$\int_{t_e}^{t_e+\mathrm{d}t_e} \frac{c\mathrm{d}t}{R(t)} = \int_{t_0}^{t_0+\mathrm{d}t_0} \frac{c\mathrm{d}t}{R(t)},$$

or

$$\frac{\mathrm{d}t_e}{R(t_e)} = \frac{\mathrm{d}t_0}{R(t_0)} \tag{7.15}$$

(assuming $\mathrm{d}t_e/t_e$, $\mathrm{d}t_0/t_0 \ll 1$). Now suppose the two events of emission correspond to consecutive wave crests (Fig. 7.2). Then

$$\frac{\nu_e}{\nu_0} = \frac{\mathrm{d}t_0}{\mathrm{d}t_e} = \frac{R(t_0)}{R(t_e)} = 1 + z \tag{7.16}$$

If the universe has expanded so that $R(t_0) > R(t_e)$, then there is a redshift ($z > 0$).

If we could observe light which was emitted from a source at $t_e = 0$, so $R(t_e) = 0$, it would be redshifted to infinite wavelength. However, as the universe is opaque for $R(t)/R(t_0) < 10^{-3}$ (Section 5.2), we know that we can observe no sources with redshift greater than 1000.

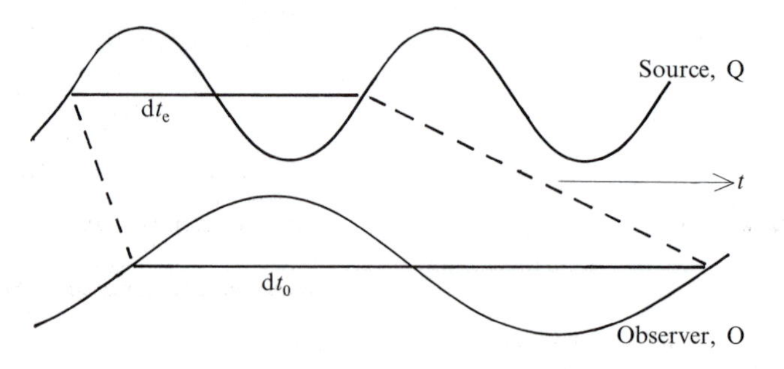

Fig. 7.2 The redshifting of a light wave from a distant source.

7.5 Luminosity distance

To calculate the flux from a distance source Q, consider a spherical surface $r = r_0$ centred on Q, passing through the observer O. Then the element of area at O defined by the four points (θ, ϕ), $(\theta + \mathrm{d}\theta, \phi)$, $(\theta, \phi + \mathrm{d}\phi)$, $(\theta + \mathrm{d}\theta, \phi + \mathrm{d}\phi)$, will subtend a solid angle

$$\mathrm{d}\Omega = \sin\theta \, \mathrm{d}\theta \, \mathrm{d}\phi$$

at Q (see Fig. 7.3). To calculate the area of this element we note that the 'proper' distance (i.e. as determined by radar methods) between two events is given by

$$(-c^2 \mathrm{d}s^2)^{1/2}|_{\mathrm{d}t=0},$$

so the area of the element is

$$R(t_0) r_0 \mathrm{d}\theta \cdot R(t_0) r_0 \sin\theta \, \mathrm{d}\phi = R_0^2 r_0^2 \mathrm{d}\Omega. \tag{7.17}$$

For a unit area,

$$\mathrm{d}\Omega = (R_0^2 r_0^2)^{-1}.$$

The energy emitted per second into $\mathrm{d}\Omega$ is $P\mathrm{d}\Omega$, so the flux received by O per unit area is

$$S = \frac{P}{R_0^2 r_0^2} Z^{-2}, \tag{7.18}$$

where one factor, Z^{-1}, is needed because the photons arrive with less energy than

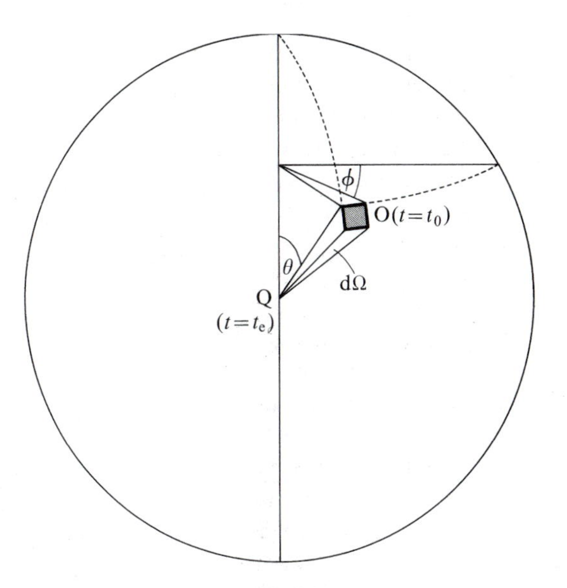

Fig. 7.3 The light emitted into a small element of solid angle $\mathrm{d}\Omega$ is received by O on the small rectangular element of area which forms part of the sphere $r = r_0$ centred on Q.

they set out with (since $E = h\nu$) and the second factor, Z^{-1}, because the photons arrive less frequently than they set off, by eqn (7.15).

From the definition of luminosity–distance (eqn (3.1)), eqn (7.18) implies

$$D_{\text{lum}} = R_0 r_0 Z, \tag{7.19}$$

where r_0 is related to the redshift by eqns (7.13) and (7.16). Eqn (7.13) can be written:

$$\int_{R_0/Z}^{R_0} \frac{c\,\mathrm{d}R}{R\dot{R}} = \begin{cases} \sin^{-1} r_0, & \text{if } k = 1, \\ r_0 & \text{if } k = 0, \\ \sinh^{-1} r_0, & \text{if } k = -1 \end{cases} \tag{7.20}$$

It can then be shown that if $\Lambda = 0$ (see Section 8.2),

$$D_{\text{lum}} = c\tau_0\{2[\Omega_0 z + (\Omega_O - 2)(\sqrt{1 + \Omega_0 z} - 1)]/\Omega_0^2\}. \tag{7.21}$$

To the first order in z we simply have the Hubble law

$$D_{\text{lum}} = c\tau_0 z$$

but the full expression (7.21) depends on the cosmological density parameter Ω_0, which we can therefore hope to determine by observing $D_{\text{lum}}(z)$.

In terms of magnitudes (Section 3.2),

$$\begin{aligned} m &= M + 5 \lg \frac{D_{\text{lum}}(z)}{10 \text{ pc}} \\ &= M + 5 \lg (c\tau_0/10 \text{ pc}) + 5 \lg \frac{2[\Omega_0 z + (\Omega_0 - 2)(\sqrt{1 + \Omega_0 z} - 1)]}{\Omega_0^2}. \end{aligned} \tag{7.22}$$

7.6 The *K*-correction

In the discussion above P and S represent the total energy per second emitted by the source and the total energy per second per unit area received by an observer. In practice we are usually observing in some relatively narrow band of wavelengths. Let $P(\nu_e)\mathrm{d}\nu_e$ be the energy emitted per second in the frequency range $(\nu_e, \nu_e + \mathrm{d}\nu_e)$ and suppose the corresponding energy received at the top of the Earth's atmosphere is $S(\nu_0)\mathrm{d}\nu_0$, where $\nu_e = Z\nu_0$. Then by eqn (7.18)

$$S(\nu_0)\mathrm{d}\nu_0 = P(\nu_e)\mathrm{d}\nu_e/D_{\text{lum}}^2$$

where D_{lum} is defined by eqn (7.19), so

$$S(\nu_0) = P(\nu_0 Z)Z/D_{\text{lum}}^2. \tag{7.23}$$

Now suppose that the Earth's atmosphere, the telescope, and the detecting system result in a fraction $\phi(\nu_0)$ of the energy incident on the atmosphere at frequency ν_0 being recorded.

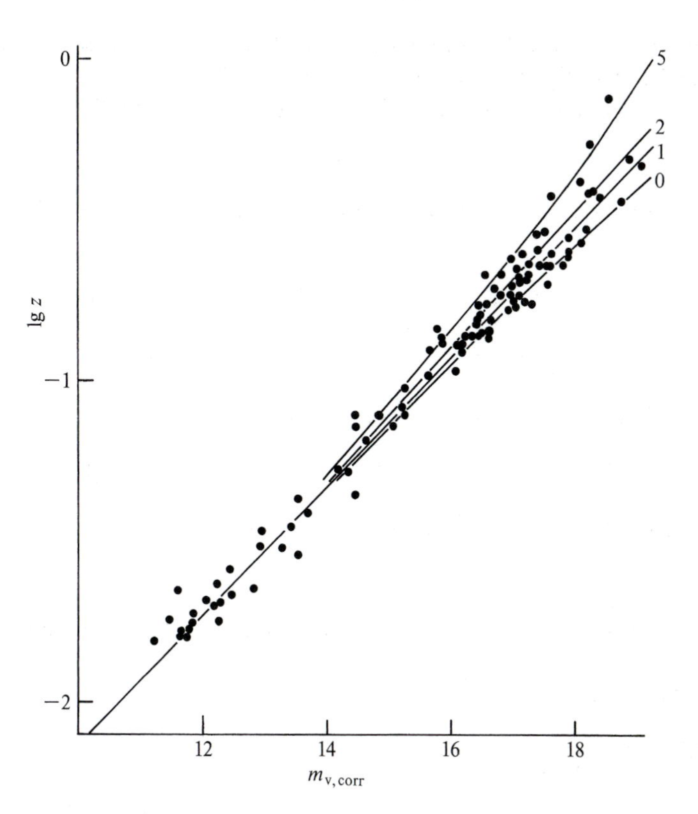

Fig. 7.4 Redshift–magnitude curves for cosmological models with different values of q_0, compared with Sandage and Visvanathan's data for the brightest galaxies in 103 clusters. The maximum redshift of Hubble's 1927 observations was 0.003. The horizontal axis is $m_{\mathrm{V.corr}} = m_{\mathrm{V}} - K(z)$.

The measured flux in the band B. say, is then

$$\begin{aligned} S_{\mathrm{B}} &= \int_{\mathrm{B}} \phi(\nu_0)\, S(\nu_0)\, \mathrm{d}\nu_0 \\ &= \frac{Z}{D_{\mathrm{lum}}^2} \int_{\mathrm{B}} \phi(\nu_0)\, P(\nu_0 Z)\, \mathrm{d}\nu_0. \end{aligned} \tag{7.24}$$

In terms of magnitudes:

$$m = M + 5 \lg \{D_{\mathrm{lum}}(z)/10\ \mathrm{pc}\} + K(z), \tag{7.25}$$

where $$K(z) = -2.5 \lg \left\{Z \int_{\mathrm{B}} \phi(\nu_0) P(\nu_0 Z) \mathrm{d}\nu_0 \Big/ \int_{\mathrm{B}} \phi(\nu_0) P(\nu_0)\, \mathrm{d}\nu_0\right\}. \tag{7.26}$$

The *K*-correction has been calculated for the U, B, V photometric bands for different types of galaxy from visible and ultraviolet observations of nearby galaxies. At radio wavelengths, or for quasars in the visible band, it is often a

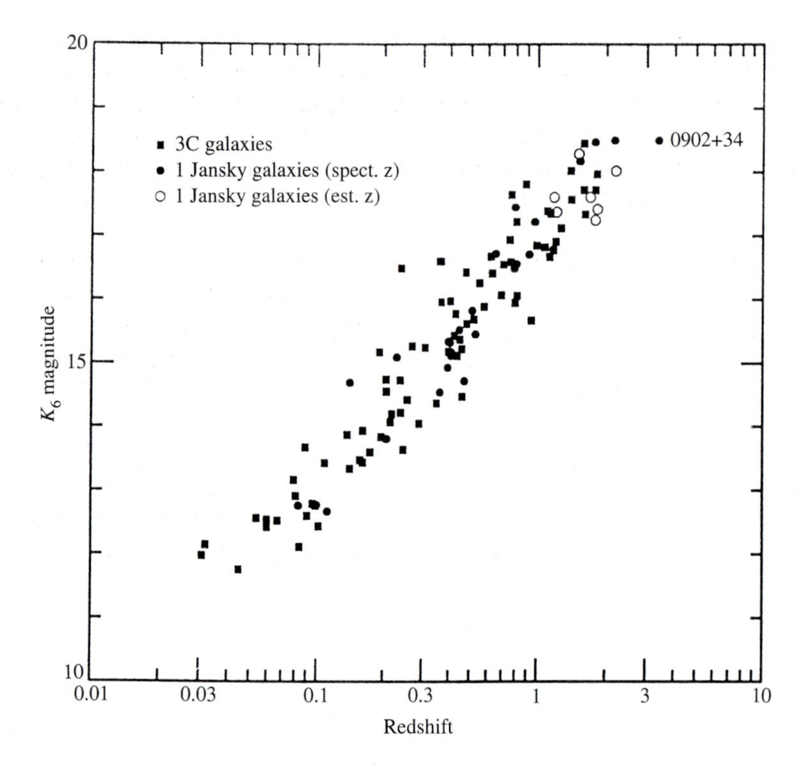

Fig. 7.5 Infrared K-magnitude versus redshift for radio galaxies.

good approximation to assume a power-law spectrum $P(\nu) \propto \nu^{-\alpha}$ and then eqn (7.23) becomes

$$S(\nu_0) = P(\nu_0)Z^{1-\alpha}/D_{\text{lum}}^2 \tag{7.27}$$

and eqn (7.26) becomes

$$K(z) = 2.5(\alpha - 1)\,\lg(1+z), \tag{7.28}$$

since $Z = 1 + z$.

Exact $m-z$ curves for different values of $q_0 = \Omega_0/2$ are shown in Fig. 7.4 compared with the observed magnitude–redshift relation for the brightest galaxies in 103 clusters. The best value of q_0 is

$$q_0 = 1.6 \pm 0.4\,. \tag{7.29}$$

However, the true uncertainty is far larger since galaxies are probably changing their luminosity with time. If galaxies were more luminous in the past (due to a more rapid formation rate of bright stars, for example) then a smaller value of q_0 is appropriate, and vice versa. One important effect is galaxy cannibalism, where the most massive galaxies in a cluster grow by merging with other galaxies.

The scatter of the points in Fig. 7.4 about the mean line arises because these galaxies do not have exactly the same luminosity. If we plot *all* galaxies in such a diagram the scatter is enormous, since the absolute magnitude of galaxies ranges over at least 12 magnitudes (a factor of 100 000 in the optical luminosity P). The same is true for quasars, assuming their redshifts are cosmological. Clearly we cannot use the magnitude–redshift diagram for all galaxies, or for quasars, to determine Ω_0.

However, another class of objects that has been used to apply this test is the radio galaxies. By taking deep photographic plates at the positions of radio sources, some very faint galaxies have been found, some of which have redshifts appreciably larger than the most distant cluster in Fig. 7.4. Figure 7.5 shows the infrared K-magnitude versus redshift for radio galaxies.

7.7 Diameter distance

Consider an object of size l at distance $r = r_0$, subtending an angle $\delta\theta$ at the origin (Fig. 7.6). From the metric (eqn (7.11)), the proper distance between the ends of the object is

$$R(t_e) r_0 \delta\theta = l, \tag{7.30}$$

by definition, so

$$\delta\theta = \frac{lZ}{R_0 r_0}. \tag{7.31}$$

From eqn (3.5), the diameter distance is then

$$D_{\text{diam}} = R_0 r_0 Z^{-1}.$$

This has been applied to bright galaxies in clusters (Fig. 7.7); the quoted value of q_0 is 0.15 ± 0.3. This test can also be applied to rich clusters of galaxies, which seem to have a core of well-defined linear size (Fig. 7.8), with the formal result $q_0 \cong -0.3 \pm 0.6$. However, this can be affected by dynamical evolution of clusters.

Notice that all the theoretical curves, except that for $q_0 = 0$, go through a minimum in $\delta\theta$, after which $\delta\theta$ starts to increase with redshift. It would be an important test of these models actually to see this happening. Since the minimum occurs at fairly large redshift, unless q_0 is unreasonably large, the best hope for testing this lies in the quasars. Figure 7.9 shows a plot of the average radio size of compact radio sources in quasars and radio-galaxies against redshift. There is

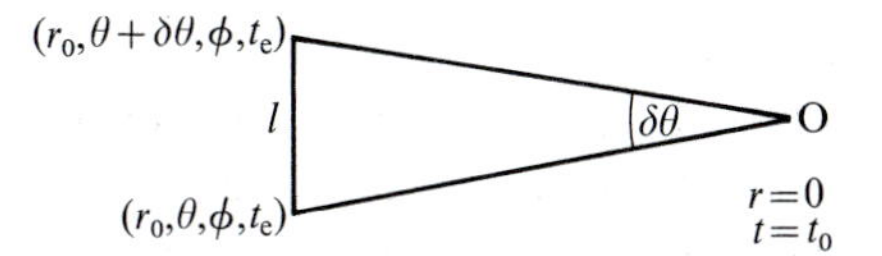

Fig. 7.6 An object of size l subtending an angle $\delta\theta$ at the observer O.

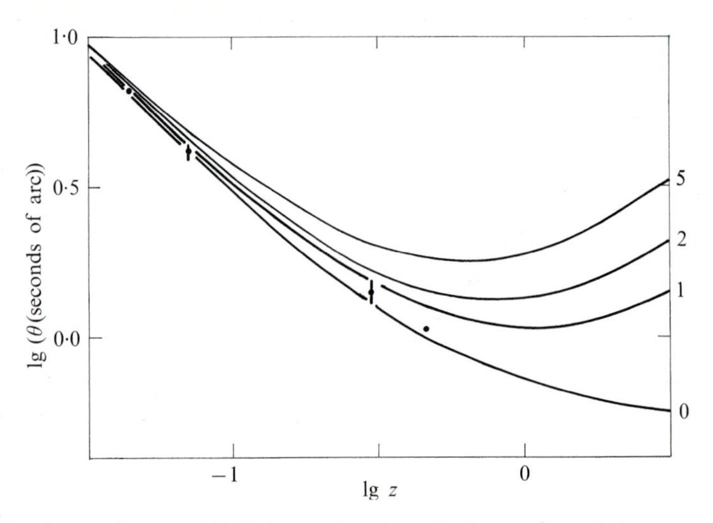

Fig. 7.7 Angular diameter–redshift diagram for galaxies in clusters. Theoretical curves are labelled with the value of q_0.

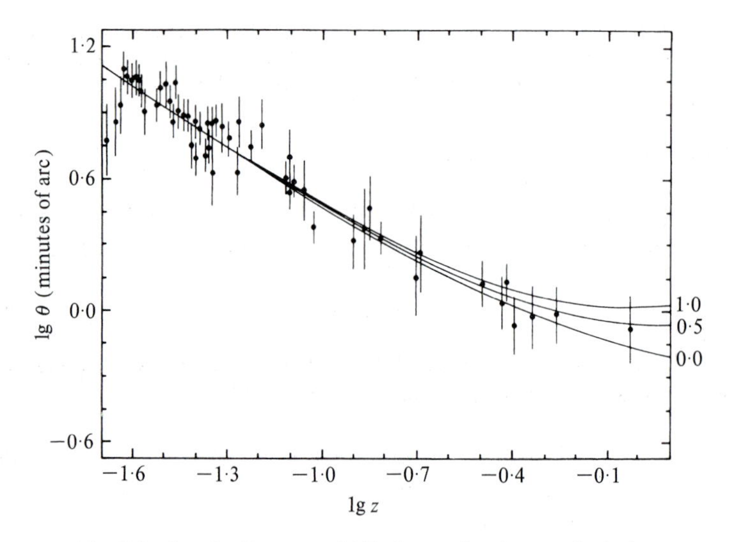

Fig. 7.8 Angular diameter–redshift diagram for clusters of galaxies.

remarkably good agreement with the theoretical curve (eqn 7.25) for $\Omega_0 = 1$, and there does seem to be evidence for the expected minimum in the $\theta - z$ relation. However, the possibility of evolution in the radio-source population makes this an uncertain way of determining Ω_0.

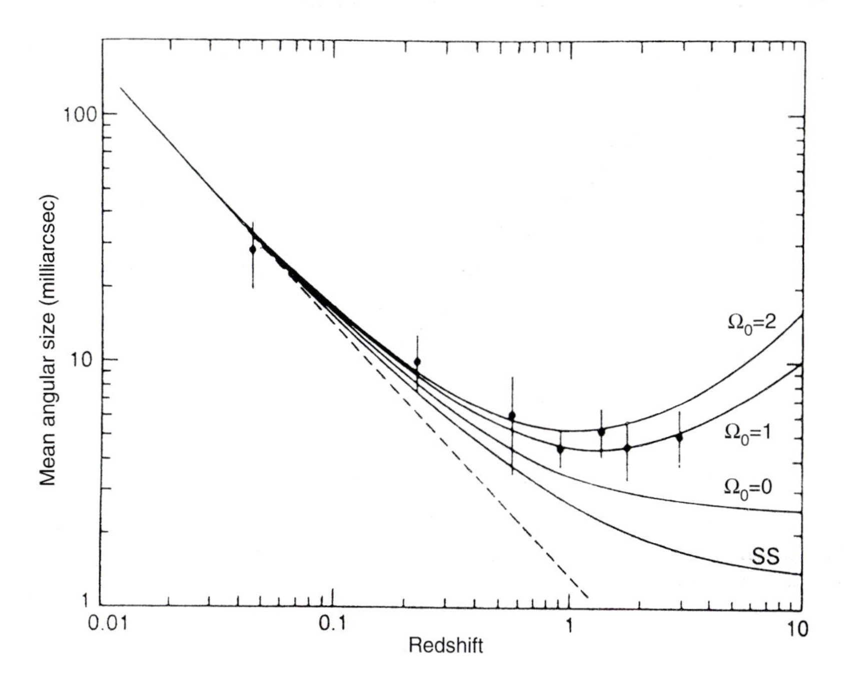

Fig. 7.9 Mean angular size against redshift for 82 compact radio sources, compared with $\Lambda = 0$ models for a standard source of size 41 pc and $\Omega_0 = 0, 1, 2$. A curve for the steady state model (see Chapter 8) is also shown. Figure from Kellermann (1993), *Nature* **361**, 134.

7.8 Number counts of sources

Consider a population of sources uniformly and randomly distributed through the matter in the universe, like the currants in a pudding. The number density of sources

$$\eta(t) \propto \rho(t) \tag{7.32}$$

provided the probability of a piece of matter being a source is independent of t. The proper volume of the element at Q is (Fig. 7.10).

$$dV = \frac{R\,dr}{(1 - kr^2)^{1/2}} \cdot Rr\,d\theta \cdot Rr\sin\theta\,d\phi \tag{7.33}$$

and for 1 sr of a spherical shell

$$dV = \frac{R^3(t)r^2\,dr}{(1 - kr^2)^{1/2}}.$$

The number of sources in 1 sr of this shell is then

$$\frac{\eta(t)R^3(t)r^2 dr}{(1 - kr^2)^{1/2}} = \frac{\eta_0 R_0^3 r^2 dr}{(1 - kr^2)^{1/2}}$$

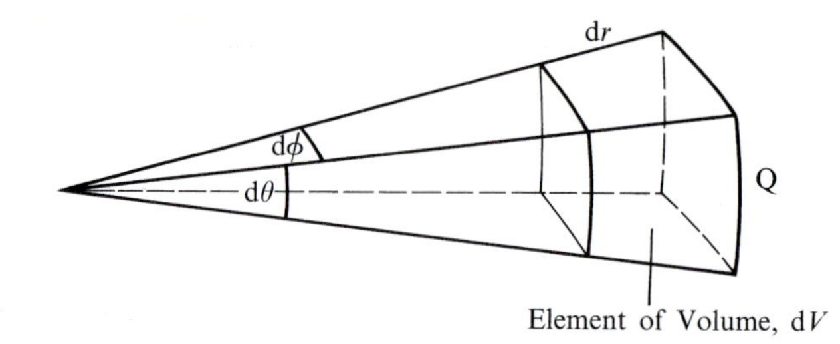

Fig. 7.10 A small element of volume.

by eqns (4.11) and (7.32), where $\eta_0 = \eta(t_0)$.

The total number of sources per steradian out to $r = r_0$ is

$$N(r_0) = \eta_0 R_0^3 \int_0^{r_0} \frac{r^2 \mathrm{d}r}{(1 - kr^2)^{1/2}}. \tag{7.34}$$

This can then be combined with eqn (7.18) to give $N(S)$, the number of sources per steradian that are brighter than S, assuming all sources have the same luminosity.

It is found that

$$\frac{\mathrm{d}\lg N}{\mathrm{d}\lg S} > -1.5 \quad \text{for all } S, q_0.$$

Figure 7.11 shows optical counts of galaxies. The observational uncertainties and an ignorance of galaxy evolution make the determination of q_0 impossible.

Counts of radio sources (Fig. 7.12) are appreciably steeper than any of the theoretical curves, implying that strong evolutionary effects must be present. The sources (mostly quasars and radio galaxies) must either have been more luminous in the past, or the probability of a source being alight must have been greater.

Optical and near-infrared counts of normal galaxies

Deep B-band counts of galaxies are too steep at faint magnitudes for passively evolving models, which allow only for the evolution of the stars in the galaxies. The redshift distribution for deep galaxy samples fits the non-evolving model and there is no tail to high redshift as would be expected for luminosity evolution, so we need either density evolution due to mergers or a new population of rapidly evolving dwarf galaxies. This faint galaxy excess is not seen in the infrared K-band, so the galaxies contributing to the excess must be bluer than average.

The infrared I-band counts also show an excess and the absence of a tail at high redshift to the z-distribution supports the rapidly evolving dwarf picture if the cosmological constant (see Chapter 8) $\Lambda = 0$. Alternatively a model with $\Omega_0 = 0.1, \lambda_0 = \Lambda/3H_0^2 = 0.9$ with no evolution could fit the counts.

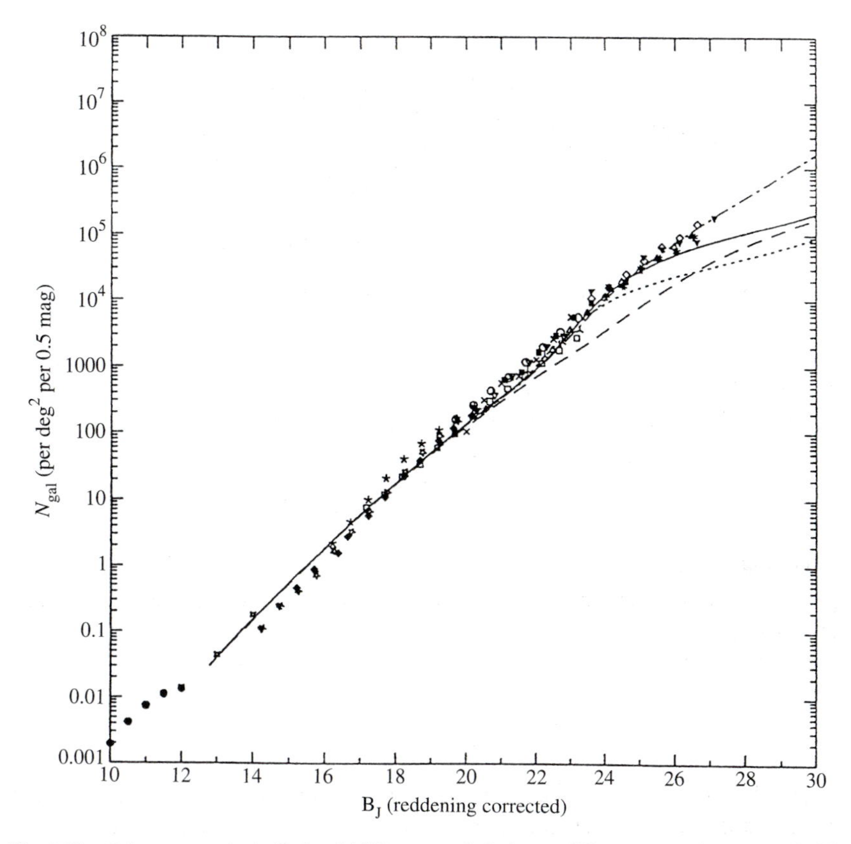

Fig. 7.11 Galaxy counts in the B_J band (different symbols denote different surveys), compared with a no-evolution model with $\Omega_0 = 0.1$ (broken curve), and with models allowing for stellar evolution in galaxies and $\Omega_0 = 0.1$ (solid curve) and 1 (dotted curve). The dash-dotted curve is the same as the latter except that the luminosity function is steepened at the faint end for $z > 1$. Figure from T. Shanks.

Radio, optical, and X-ray counts of active galaxies

Radio source counts imply strong evolution of radio galaxies and radio-loud quasars. The evolution is close to a pure 'luminosity' evolution, in which the luminosity function satisfies

$$\phi(L, z) = \phi\{L, L^*(z)\} \tag{7.35}$$

with

$$L^*(z) \propto \begin{array}{ll} (1+z)^{3.1}, & 0 < z < 2, \\ = \text{constant} & 2 < z < 5. \end{array}$$

Optical surveys of (predominantly radio-quiet) quasars also show strong evolution, which is also close to a pure luminosity evolution of the same form.

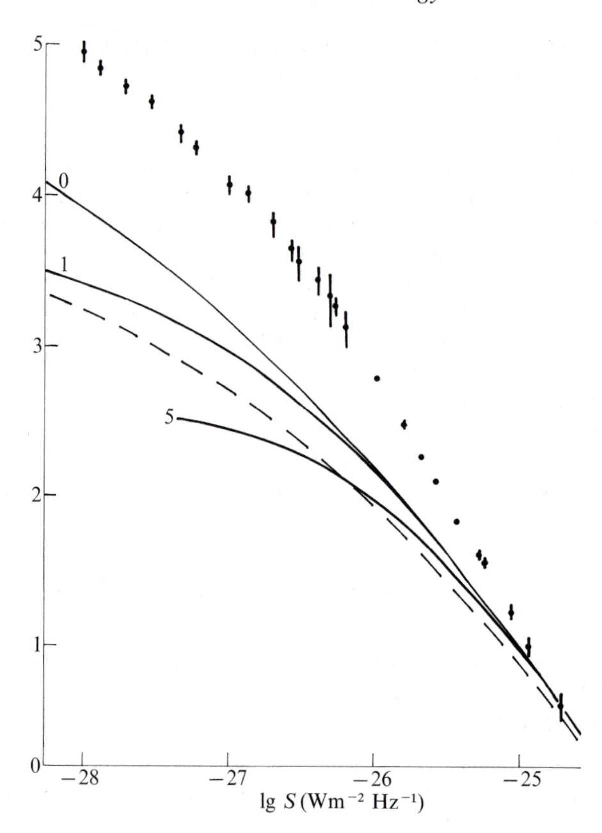

Fig. 7.12 Counts of radio sources, compared with theoretical curves for different q_0 (solid curves) and for the steady state model (broken curve—see Section 8.3), showing strong evolutionary effects in the radio-source population.

Deep X-ray counts made with the ROSAT satellite are dominated by quasars and also show strong evolution.

Far-infrared and radio counts of starbust galaxies

Redshift surveys of IRAS 60 μm galaxies show clear evidence for strong evolution at a similar rate to that seen in quasars and radio galaxies. Source counts at 60 μm also show strong evidence for evolution.

Because of the strong correlation between far-infrared and radio luminosity for spiral galaxies, sub-mJy radio source counts at 1.4 GHz and redshift surveys of faint samples of radio sources can provide powerful constraints on the evolution of the starbust galaxy population. Below 1 mJy the dominant population of radio-sources is starbust galaxies and they seem to be undergoing strong evolution at a rate similar to eqn (7.35).

Table 7.1 Some of the great catalogues of modern astronomy

Optical

1. *The reference catalogue of bright galaxies*

Compiled by G. and A. de Vaucouleurs, this comprises essentially the galaxies from the New General and Index Catalogues (NGC† and IC†) of nebulae, in turn based on the General Catalogue of William and John Herschel. It contains information on 2599 galaxies, and is complete to about 13th magnitude.

2. *The Zwicky catalogue of galaxies and clusters of galaxies*

Compiled by F. Zwicky and coworkers, it contains 31 000 galaxies in the northern hemisphere brighter than $m_{pg} = 15.7$, and lists 9700 clusters of galaxies.

3. *The Abell catalogue of rich clusters of galaxies*

Lists 2700 clusters rich in galaxies

4. Catalogues of unusual objects

(a) *Zwicky lists of compact galaxies*; (b) *Arp catalogue of peculiar galaxies*; (c) *Vorontsou–Velyaminov catalogue of interacting galaxies*; (d) *Markarian catalogue of blue galaxies*.

Infrared†

IRAS Point Source Catalog of sources detected at 12, 25, 60, and 100 μm

Radio†

Revised 3rd Cambridge catalogue of northern hemisphere sources brighter than 9 Jansky $\{1\ \text{Jy} = 10^{-26}\ \text{Wm}^{-2}\ \text{Hz}^{-1}\}$ *at* 178 *MHz*.

Parkes catalogue of southern hemisphere sources.

X-ray†

The 4th *Uhuru catalogue* of bright sources in the range 2–8 keV, compiled from data from the Uhuru satellite.

---

†Contains some Galactic sources also.

The most likely mechanism for driving the evolution in these different types of active galaxy is interactions between galaxies or mergers, which can concentrate gas towards the nuclei of galaxies, feeding a black hole there, and causing a burst of star formation.

7.9 The luminosity–volume test

These evolutionary effects can also be seen vividly by means of the *luminosity–volume* test, which combines the luminosity–distance and source-count tests into a single, more powerful test, provided a complete sample of sources down to some limiting flux level S_{min} is available. For a source in the sample with flux S and redshift z we can calculate the luminosity P from eqn (7.18) for some chosen cosmological model. Sources of this luminosity P should then be uniformly

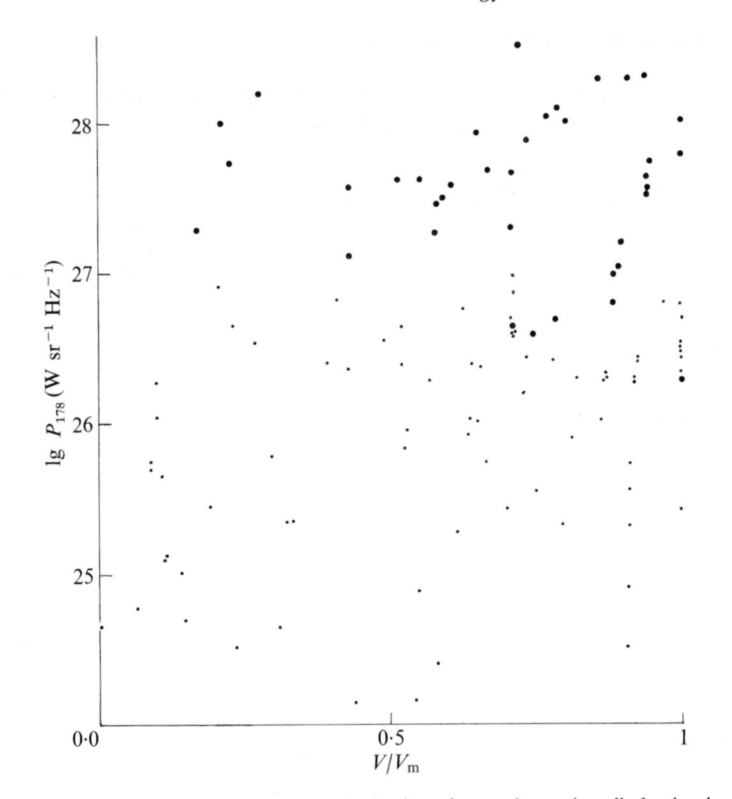

Fig. 7.13 The luminosity–volume test. The distribution of monochromatic radio luminosity (at 178 MHz) against V/V_m for all quasars (large dots) and radio galaxies (small dots) in the revised third Cambridge catalogue of radio sources (see Table 7.1) with visual magnitude brighter than 19.5, and with $|b| > 10°$, $\delta > 10°$, where b is the Galactic latitude and δ is the declination. The distribution for quasars and for the more luminous radio galaxies is non-uniform, showing that these populations have changed their properties dramatically with time. The calculations have been performed in the Milne model ($q_0 = 0$).

distributed with respect to the co-moving volume, $V(z)$ (i.e. when the effect of the expansion of the universe is allowed for). Of course, there will be a redshift z_{max} at which a source of this luminosity P would disappear out of the sample (its flux would drop below S_{min} for $z > z_{max}$ so in fact we can test this uniformity only for $0 < V(z) \leq V(z_{max}) = V_m$, say. In practice it is best to calculate V/V_m for each source in the sample and then look at the distribution of luminosity with respect to V/V_m. Figure 7.13 shows such distributions for radio galaxies and for quasars in the third Cambridge catalogue of radio sources (see Table 7.1; p. 125). The distributions for quasars and for the more luminous radio galaxies are strikingly non-uniform (more sources are found at large values of V/V_m), showing that evolutionary effects are present. Similar results are obtained for other values of q_0.

7.10 Integrated background radiation

The intensity of the integrated background radiation from a population of sources of luminosity P, and number density at the present epoch η_0, is

$$I = \int_{S=0}^{\infty} S \, \mathrm{d}N(S) \quad \mathrm{W\,m^{-2}\,sr^{-1}} \tag{7.36}$$

$$= \eta_0 P R_0 \int_{Z=1}^{x} \frac{\mathrm{d}r}{Z^2(1-kr^2)^{1/2}} = \eta_0 P c \int_{t=0}^{t_0} \frac{R(t)}{R_0} \mathrm{d}t. \tag{7.37}$$

Using eqn (7.12):

$$I < \eta_0 P c \tau_0$$

for all models.

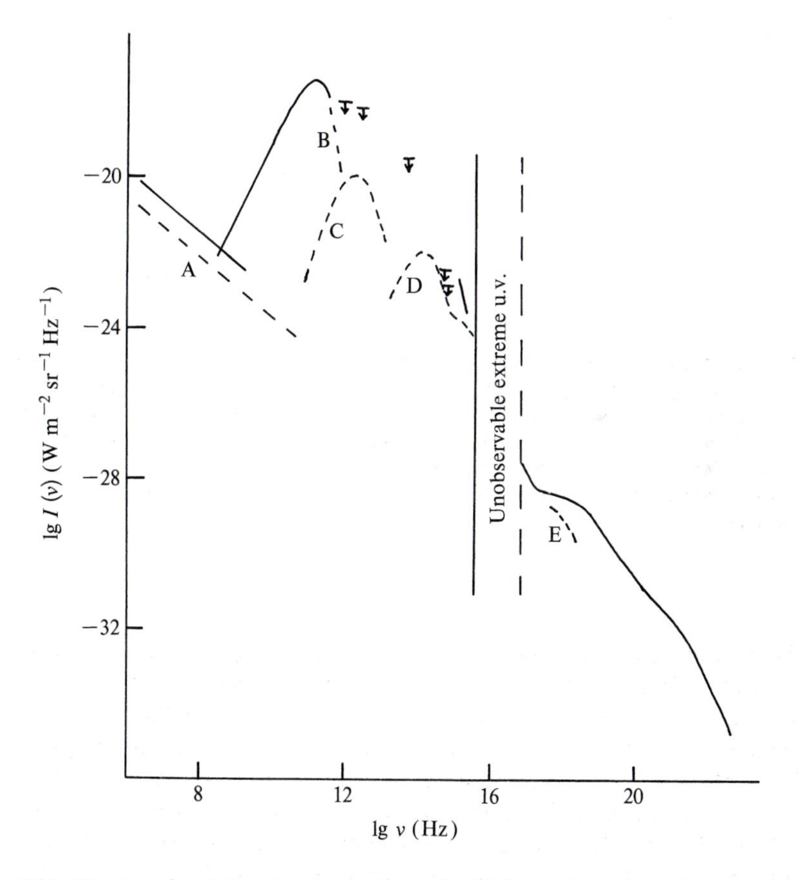

Fig. 7.14 The intensity of the integrated background radiation. Solid curves and upper limits: observations. Broken curves: predictions assuming no evolution. A, radio galaxies and quasars; B, 2.7 K cosmic black body; C, radiation from dust in galaxies; D, integrated starlight in galaxies; E, rich clusters.

If we now observe the intensity of the background in the frequency range $(\nu_0, \nu_0 + d\nu_0)$ to be $I(\nu_0)d\nu_0$, then

$$I(\nu_0) = \eta_0 c \int_{t=0}^{t_0} P(\nu_0 Z)\, dt \tag{7.38}$$

where $Z = R_0/R(t)$.

Figure 7.14 shows the predicted background, using the Milne model, as a function of frequency for different populations of source, compared with the observed background.

Radio background

The main contribution is from the Milky Way, but it can be shown that the steep-spectrum (mean spectral index $\alpha \sim 0.8$—Section 1.4; p. 16) extragalactic sources contribute an intensity of about $10^{-21.6}$ W m^{-2} Hz^{-1} sr^{-1} at 178 MHz, about 10 times more than that predicted assuming no evolution. This confirms that evolution has to be present, as was suggested by the source counts.

Microwave and infrared background

The dominant observed background in the microwave region is the 2.7 K blackbody radiation. Due to the foreground radiation from the Earth's atmosphere and the emission from interstellar and interplanetary dust, we have only upper limits at the moment on the infrared background. The predicted contribution from interstellar dust in galaxies and from starburst galaxies is shown in Fig. 7.14.

Optical background

The predicted integrated background from the starlight in galaxies falls well below the observational limits. The problem is that the brightness of the light from the Earth's atmosphere (for ground-based observations), from zodiacal light, and from the starlight of the Milky Way together swamp out the cosmic background. Possible changes with time in the luminosities of galaxies due to stellar evolution lead to uncertainties in the theoretical curves.

Ultraviolet and soft X-ray background

Observations from satellites and rockets show the existence of background radiation emission concentrated towards the Galactic plane, probably due to starlight scattered by dust in the ultraviolet and to emission from hot (10^5–10^6 K) interstellar gas at soft X-ray wavelengths. At high Galactic latitudes an isotropic component can be seen which might be due to a hot intergalactic medium ($\sim 10^6$ K) with the closure density ($\Omega = 1$). However, the ultraviolet background at high Galactic latitudes may well be scattered starlight and the soft X-ray background is probably due to a halo of hot gas surrounding our Galaxy. The extragalactic background in the wavelength range 100–912 Å is unobservable due to interstellar absorption.

Hard X-ray and gamma-ray background

The Einstein satellite has found that the largest contribution to the 1–3 keV background is from quasars. Other classes of discrete source which are known to make a significant contribution to the 2–10 keV background are clusters of galaxies (~10 per cent) and emission line galaxies (~30 per cent). These results have been confirmed by the ROSAT mission.

Over the range 2–130 keV the background can be fitted by thermal Bremsstrahlung from a hot (4.4×10^8 K) intergalactic gas ($\Omega = 0.46$), but there is considerable difficulty in accounting for the energy to heat this gas.

An additional contribution to the background will be inverse Compton radiation (Section 1.4; p. 16) from normal galaxies, from radio galaxies, and from quasars, due to the interaction of the relativistic electrons responsible for their radio synchrotron radiation with the photons of the blackbody radiation. However, unless the magnetic fields in these objects are surprisingly low, the contribution to the background is likely to small.

The γ-ray background shows a considerable enhancement towards the Galactic plane, and this is believed to be due to the decay of π^0-mesons created by high-energy cosmic rays ploughing through the interstellar gas of our Galaxy. The

Table 7.2 Properties of selected cosmological models

Quantity	Milne	E–de S	de S	Steady state
k	-1	0	0	0
$\lambda = \Lambda/3H_0^2$	0	0	1	1
$R(t)$	ct	$R_0(t/t_0)^{\frac{2}{3}}$	$R_0 \exp(t - t_0)/\tau_0$	
$\tau_0(= 1/H_0)$	t_0	$3t_0/2$	constant	constant
q_0	0	$\frac{1}{2}$	-1	-1
Ω_0	0	1	0	$\frac{1}{2}$
R_0	$c\tau_0$	$c\tau_0$	$c\tau_0$	$c\tau_0$
$r_0(z)$	$(Z - 1/Z)/2$	$2(1 - Z^{-\frac{1}{2}})$	z	z
$D_{\text{lum}}(z) = R_0 r_0(z) Z$				
$D_{\text{diam}}(z) = R_0 r_0(z)/Z$				
$N(z)/n_0(c\tau_0)^3$	$(Z^2 - Z^{-2})/8 - \ln Z/2$	$8(1 - Z^{-\frac{1}{2}})^3/3$	$z^3/3$	$\ln Z - (2z + 3z^2)/2Z^2$
$I(\nu_0)/\eta_0 c\tau_0 P(\nu_0)$	$1/(1+\alpha)$	$2/(3 + 2\alpha)$	$1/\alpha$	$1/(3 + \alpha)$

Note: $Z = 1 + z$

main contribution to the isotropic component of the background is probably from quasars and Seyfert galaxies. More exotic possibilities are Hawking radiation from primordial black holes) and π^0-decay due to matter–antimatter annihilation (Epilogue, p. 149).

7.11 Problems

7.1 Work out the detailed predictions of the tests described in this chapter for the cases $q_0 = 0, 1/2$.

7.2 For a universe with $k = 0$, $R(t) = R_0(t/t_0)^n$, where $n < 1$, use eqns (7.16) and (7.20) to deduce that

$$r_0 = (ct_0/(1-n)R_0)\,\{1-(1+z)^{1-1/n}\}.$$

For $n = 2/3$, deduce that the proper distance to a quasar at redshift $z = 5$ is $c\tau_0\{6-\surd 6\}/3$.

7.3 Show that for a population of sources with the same linear size, l, in the Einstein–de Sitter model there is a minimum observed angular diameter at $z = 5/4$.

8
Other cosmological theories

8.1 Introduction

So far we have described the standard general relativistic, isotropic, homogeneous, hot big-bang models and their consequences. In this chapter we look at some alternative ideas both within and outside the framework of general relativity.

First we look at the consequences of including a non-zero 'cosmological' term in the field equations of general relativity. For historical reasons we give an account of the steady state theory. We look at some theories of gravity in which the gravitational 'constant' G varies with time, and at anisotropic and inhomogeneous models within general relativity. We mention the cosmological implications of 'grand unified' theories, which attempt to unify the strong, electromagnetic, and weak forces of physics, and we discuss models which dispense with a hot big bang altogether. Finally we look into Eddington's 'magic' numbers and the 'anthropic' principle.

8.2 General relativistic models with the Λ-term

When Einstein originally put forward his general theory of relativity, he included an additional term in the field equations, the so-called *cosmological term.* This modifies the law of gravitation at large distances into an attraction or repulsion directly proportional to distance, $\ddot{\boldsymbol{r}} = \Lambda \boldsymbol{r}/3$, Λ constant. No such effect is observed in the solar system, or in the structure of our Galaxy, so Λ must be very small. This extra term would have an effect only on the scale of clusters of galaxies or larger, hence the name for Λ of *cosmological constant.*

The Λ-term is consistent with all the basic principles that led Einstein to his field equations (it is effectively a constant of integration), but it is usually set equal to zero by relativists in order to keep the theory as simple as possible. However, it leads to some new cosmological possibilities, which we will now investigate.

The equations for the scale factor $R(t)$, (4.12) and (4.13) (p. 67), which are derived from the field equations assuming the cosmological principle, become

$$\ddot{R} = -4\pi G\rho_0 R_0^3/3R^2 + \Lambda R/3 \tag{8.1}$$

$$\dot{R}^2 = 8\pi G\rho_0 R_0^3/3R - kc^2 + \Lambda R^2/3 = G(R). \tag{8.2}$$

First we see that the Λ-term does not have any effect near $R = 0$, so behaviour near the 'big bang' is unaltered. We consider the cases $\Lambda < 0$ and $\Lambda > 0$ separately:

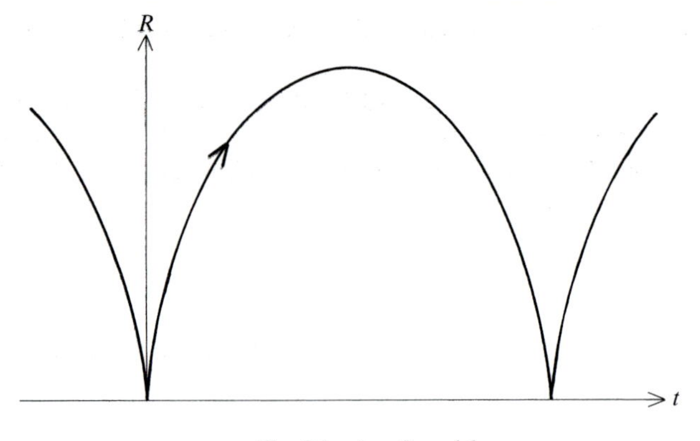

Fig. 8.1 $\Lambda < 0$ models.

$\Lambda < 0$

R has to be finite for $\dot{R}$ to remain a real number, and there exists an R_c such that $G(R_c) = 0$, i.e. $\dot{R} = 0$ when $R = R_c$. Eqn (8.1) then shows that $\ddot{R} < 0$ when $R = R_c$, so the universe starts to contract at this point. We therefore have oscillating models if $\Lambda < 0$, whatever the value of k (see Fig. 8.1).

$\Lambda > 0$

If $k \leq 0$, $\dot{R}^2 > 0$ for all R, to we have a monotonic expanding universe, the only difference from those with $\Lambda = 0$ being that at large R, $\dot{R}^2 \sim \Lambda R^2/3$, so

$$R \propto \exp\{(\Lambda/3)^{1/2} t\}. \tag{8.3}$$

If $k = 0$, $\rho_0 = 0$, $\Lambda > 0$ we have the de Sitter model, for which eqn (8.3) holds for all t.

If $k = 1$ there is a critical value of Λ, Λ_c, such that $\ddot{R} = 0$ and $\dot{R} = 0$ can both be satisfied simultaneously. From eqn (8.1), $\ddot{R} = 0$ implies

$$R = R_0(4\pi G\rho_0/\Lambda)^{1/3} = R_c, \quad \text{say}, \tag{8.4}$$

and then eqn (8.2) implies that

$$0 = (4\pi G\rho_0)^{2/3}\Lambda^{1/3}R_0^2 - kc^2$$

so,

$$\Lambda_c = (kc^2)^3/R_0^6(4\pi G\rho_0)^2. \tag{8.5}$$

This means that there is the possibility of a static model of the universe, with $R = R_c$, $\Lambda = \Lambda_c$, for all time t, provided

$$\Lambda = 4\pi G\rho_c = kc^2/R_c^2 \tag{8.6}$$

and since $\rho_c > 0$, k must be positive for this to happen. This is the Einstein static model, the first solution of general relativity to be found that satisfies the cosmological principle.

By studying the function $G(R)$ (eqn (8.2)) as a function of R, we can see what other possible $\Lambda > 0$, $k = +1$, models there are (Fig. 8.2). Clearly $G(R) \to \infty$ both for $R \to 0$ and for $R \to \infty$, and reaches a minimum at R_c, with $G(R_c) >$ or < 0 according as $\Lambda >$ or $< \Lambda_c$.

$\Lambda > \Lambda_c$

$G(R) > 0$ all R, so we have a monotonic expanding universe again

$\Lambda = \Lambda_c$.

Apart from the Einstein static model, there are two models that approach this asympotically, corresponding to the two branches of $G(R)$ — see Fig. 8.2. One expands out gradually from the Einstein state at $t = -\infty$ and then turns into an exponential expansion (eqn (8.3)). The other expands out from the usual big bang and then tends asymptotically to the Einstein model as $t \to \infty$. These are called the Eddington–Lemaître models (Fig. 8.3), EL1 and EL2.

If $\Lambda = \Lambda_c(1 + \epsilon)$, $\epsilon \ll 1$, we have the Lemaître models. For a long period of time R is close to R_c and the cosmological repulsion and gravitational attraction are almost in balance. Finally the repulsion wins and the expansion continues again.

$0 < \Lambda < \Lambda_c$

There are no solutions for $R_1 < R < R_2$ (Fig. 8.2). The solution with $R \leq R_1$ is an 'oscillating' model. In the one with $R \geq R_2$, the universe 'bounces' under the action of the cosmological repulsion (Fig. 8.4).

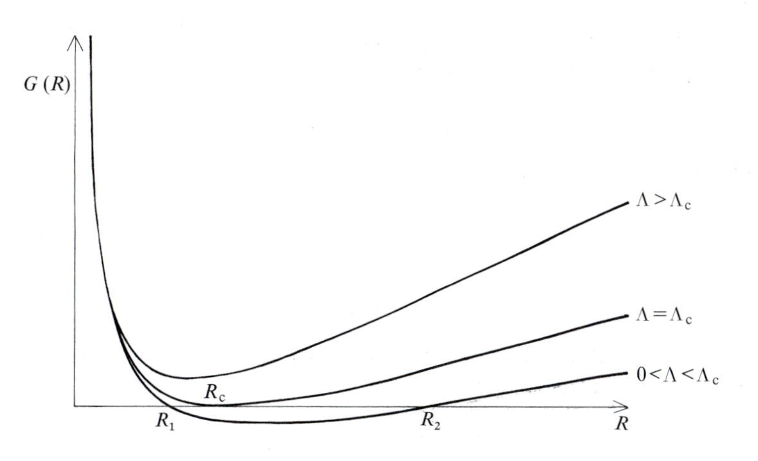

Fig. 8.2 $G(R)$ for $\Lambda > 0$, $k = +1$ models.

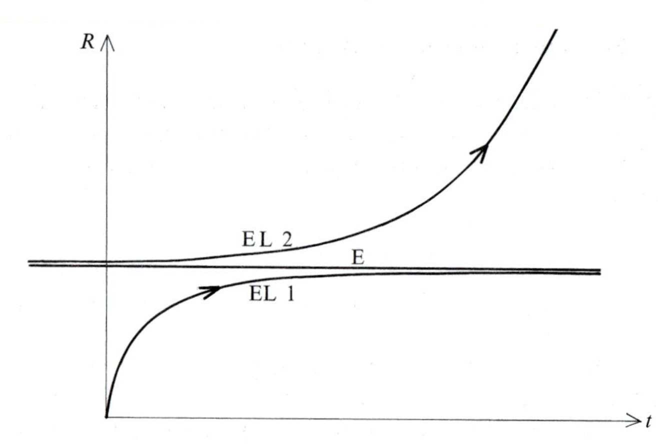

Fig. 8.3 $k = +1$, $\Lambda = \Lambda_c$ models. E = Einstein static model; EL1, EL2 = Eddington–Lemaître models.

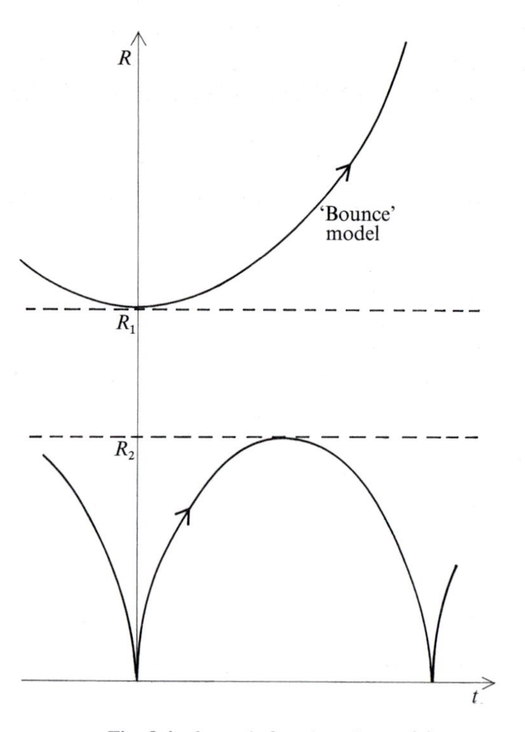

Fig. 8.4 $k = +1$, $0 < \Lambda < \Lambda_c$ models.

8.3 Observable consequences of the Λ-term

The Einstein static model can be eliminated immediately since it does not predict a redshift. There remain two models which do not originate in a big bang, the EL2 (Fig. 8.3) and 'bounce' models. In each case there would be a maximum redshift defined by

$$1 + z_{\max} = R_0/R_{\min}$$

and in the bounce models more distant objects would show a blueshift. If quasar redshifts are cosmological then $z_{\max} \gtrsim 5$.

The Lemaître models permit ages of the universe far greater than the Hubble time τ_0 (the EL2, bounce, and de Sitter models have infinite ages). Since these models have positive curvature (see Fig. 4.6, p. 72) and are spatially finite, there is the intriguing possibility of seeing all the way round the universe, and even of seeing a ghost Milky Way (normally light does not have time to make this circumnavigation, and therefore there is a horizon—Section 4.9). So far we have no evidence that this is happening. Actually, since in general relativity the large-scale topology of the universe is not specified, ghost images could in principle arise in any model. Another effect of the long 'coasting' period (Fig. 8.5) is that there would be a concentration of objects with redshifts given by

$$1 + Z \sim R_0/R_c.$$

For models with non-zero Λ, and neglecting radiation, the age of the universe can be written

$$t_0 = \int_0^{t_0} \mathrm{d}t = \int_0^{R_0} \mathrm{d}R/\dot{R}$$
$$= \tau_0 \int_0^1 \{\Omega_0/x - 3\Omega_0/2 + q_0 + 1 + (\Omega_0/2 - q_0)x\}^{-1/2},$$

where $x = R/R_0$.

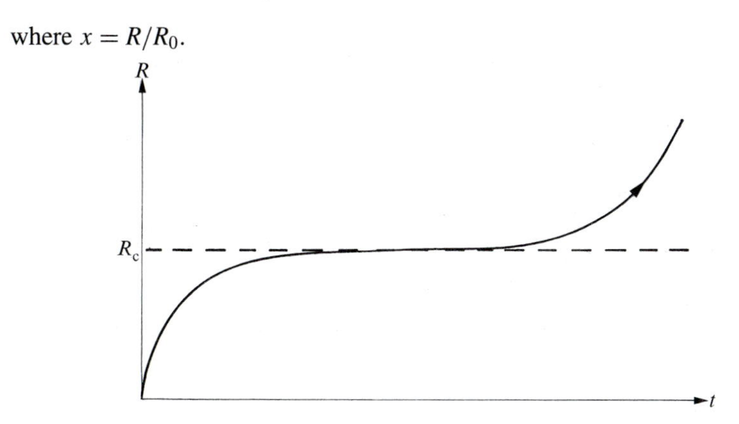

Fig. 8.5 Lemaître models, with long 'quasi-stationary' period.

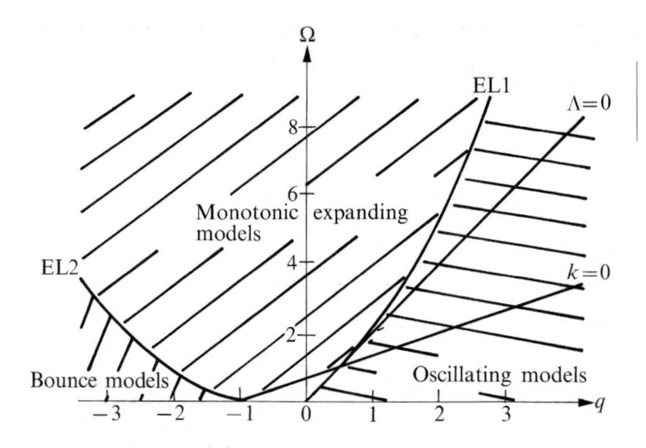

Fig. 8.6 The regions of the Ω–q plane occupied by different types of cosmological model.

If observations show that the Hubble time, τ_0, is less than the age of the universe, t_0, then we would be forced to conclude that $\Lambda > 0$. We could still retain a flat model of the universe ($k = 0$), as required by inflation, provided

$$\lambda_0 + \Omega_0 = 1,$$

where $\lambda_0 = \Lambda/3H_0^2$. For example, for $k = 0$,

$\Omega_0 = 1 - \lambda_0 =$	0.05	0.1	0.2	0.3	0.5	1
$t_0/\tau_0 =$	1.49	1.28	1.08	0.96	0.83	0.67

The best evidence for non-zero Λ would be if the parameters $\Omega/2$ and q were unequal at the present epoch, since eqns (8.1) and (4.20)–(4.22) imply that

$$\Lambda/3 = H_0^2/(\Omega_0/2 - q_0). \tag{8.7}$$

We have seen that the observed matter in galaxies corresponds to $\Omega_0 \sim 0.1$ (eqn (6.8)), whereas the magnitude–redshift test for bright galaxies gave $q_0 \sim 1$, but both these estimates are far too uncertain to be interpreted as implying $\Lambda \neq 0$.

The optical depth for gravitational lensing of distant quasars by intervening galaxies is a sensitive function of the geometry of the universe and so can provide useful constraints on the value of Λ. The present evidence from the redshift distribution of lensed quasars favours $\lambda_0 = \Lambda/3H_0^2 < 0.95$.

The condition $\Lambda = \Lambda_c$, which defines the Eddington–Lemaître models, can be shown to be equivalent to

$$(3\Omega/2 - q - 1)^3 = 27(\Omega/2)^2(\Omega/2 - q). \tag{8.8}$$

The zones of the different models in the Ω–q diagram are shown in Fig. 8.6.

8.4 The steady state cosmology

This was put forward in 1948 by Bondi, Gold, and Hoyle. The cosmological principle was strengthened to the '*perfect*' *cosmological principle*: the universe presents the same appearance to all fundamental observers *at all times*.

An immediate consequence of this is that the Hubble time must be a constant:

$$\tau = R/\dot{R}, \quad \text{constant for all } t.$$

Thus

$$R \propto \exp t/\tau, \quad \text{or} \quad R = R_0 \exp(t - t_0)/\tau. \tag{8.9}$$

A second consequence is that the density of matter is constant, and to maintain this we must have continuous creation of matter at a steady rate per unit volume

$$\frac{1}{R^3}\frac{\mathrm{d}}{\mathrm{d}t}(\rho R^3) = 3\rho/\tau. \tag{8.10}$$

It can also be shown that the three-spaces $t =$ constant have a three-dimensional curvature, known as the Gaussian curvature, of $k/R(t)^2$, which would depend on time unless $k = 0$.

We have therefore shown that the steady state metric is

$$\mathrm{d}s^2 = \mathrm{d}t^2 - \frac{\exp(2t/\tau)}{c^2}(\mathrm{d}x^2 + \mathrm{d}y^2 + \mathrm{d}z^2), \tag{8.11}$$

and this is the same as that for the de Sitter model (Section 8.1).

If field equations similar to those of general relativity are used, but with an extra term representing the creation of matter, it is found that

$$8\pi G\rho\tau^2/3 = 1, \tag{8.12}$$

which with $\tau \sim 2 \times 10^{10}$ years gives $\rho \sim 5 \times 10^{-27}$ kg m^{-3}, about 10 times higher than the density of matter in galaxies (eqn (6.10)) The remainder could be made up of ionized intergalactic hydrogen at a temperature $\sim 10^6$ K (Section 6.5). The corresponding creation rate (of, presumably, cold, neutral, uniform hydrogen) is the undetectable 10^{-44} kg m^{-3} s^1. The magnitude–redshift relation for the steady state theory is the same as for the de Sitter model:

$$m = A + 5\lg\{z(1 + z)\} \tag{8.13}$$

(where A is constant), equivalent to $q_0 = -1$, which is rather improbable from the data on bright cluster galaxies (Fig. 7.4, p. 117), bearing in mind that no evolution is permitted.

The number of sources per steradian out to redshift z is

$$\begin{aligned} N(z) &= \int_0^{r_0(z)} \eta R^3 r^2 \mathrm{d}r = \eta(c\tau)^3 \int_0^z (1 + z)^{-3} z^2 \mathrm{d}z \\ &= \eta(c\tau)^3 \left\{ \ln(1 + z) - \frac{2z + 3z^2}{2(1 + z)^2} \right\}. \end{aligned} \tag{8.14}$$

The corresponding $N(S)$ curve is flatter than -1.5, whereas the radio-source counts are appreciably steeper (Fig. 7.12, p. 124).

The final blow for the steady state theory was the isotropic 2.7 K blackbody radiation, for which no convincing explanation was produced, whereas for big-bang cosmology this is one of its greatest successes.

However, it should be remembered that in 1948, the accepted estimate of the Hubble time was only 2×10^9 years, only 10 per cent of the age of our Galaxy, Steady state cosmology provided an elegant way out of that difficulty, now resolved by new measurements of τ_0.

Hoyle and collaborators have, by analogy with some of the current inflationary scenarios, suggested a modification of the steady state theory whereby the universe is in a steady state on the large scale but some regions, including the one we find ourselves in, undergo an evolutionary phase.

8.5 Theories in which *G* changes with time

Theories of this type were first proposed by Milne, Dirac, and Jordan. More recently Brans and Dicke, Hoyle and Narlikar, and Dirac have put forward more elaborate theories of this type.

A variation of G with time has a considerable effect on the evolution of the Earth and Sun, and on the orbits of the moon and planets. If gravity has changed appreciably over the lifetime of the Earth, the radius of the Earth might have been affected. It has been suggested that the continents all fitted together at one time on a much smaller Earth. As the gravitational constant reduced, the Earth expanded to its present size and the continents were forced apart. Also a star in its hydrogen-burning phase, like the Sun, has a luminosity

$$L \propto G^7, \tag{8.15}$$

and so would have been appreciably brighter in the past if G decreases with t. The effect of this on life on Earth would be enhanced by the fact that the Earth must be moving away from the Sun if G is declining. If $G \propto t^n$, then the temperature of the Earth is $T \propto t^{9n/4}$, assuming that the Earth always absorbs the same fraction of the incident energy from the Sun. Paleoclimatologists argue that the flux of solar energy at the Earth's surface can have varied little over the Earth's history. A slightly lower solar flux results in the Earth becoming completely covered with ice, which can then only be removed by a very substantial rise in the solar flux. This argument seems to be inconsistent even with the standard ($G =$ constant) model, since the Sun is expected to have increased its luminosity noticeably over the Earth's lifetime. The existence of life on Earth three thousand million years ago limits the Earth's surface temperature to well below the boiling point of water at that time. And too fast a decline in G would lead to the Sun's having already become a red giant.

A varying G leads to a variation in the Moon's distance and period, and measurements of these lead to the most reliable limits on the variation of G obtained to date. The orbits of the planets are also modified, and this could show up in radar time-delay experiments.

The upper limit on the variation of G from these solar system arguments is now

$$|\dot{G}/G| \lesssim 3 \times 10^{-11}\ \mathrm{yr}^{-1}$$

and this is sufficient to conflict with the simplest varying G models, in which $G \propto t^{-1}$, assuming the Hubble constant $H_0 \gtrsim 5 \times 10^{-11}\ \mathrm{yr}^{-1}$. However, there are many problems of interpretation associated with these experiments and more sensitive measurements are highly desirable. It should also be emphasized that some theories in which G varies also predict other changes (to preserve energy conservation, for example) which can mask the above effects.

Even stronger limits on $\dot{G}$ follow if ^{4}He and ^{2}H are believed to be synthesized in the fireball phase of a hot big bang,

$$|\dot{G}/G| \lesssim 10^{-12}\,\mathrm{yr}^{-1}.$$

The Brans–Dicke cosmology represents the simplest extension of general relativity. In addition to the tensor gravitational field represented by the metric tensor (Section 4.4), there is a scalar field (the *gravitational 'constant'*) which is a function of time only.

Brans and Dicke take $\Lambda = 0$ and seek to satisfy *Mach's principle*, that local inertial properties should be determined by the gravitational field of the rest of the matter in the universe, by taking

$$G^{-1} \sim \sum_{\text{universe}} m/rc^2. \tag{8.16}$$

The models for the case $k = 0$ are particularly simple, since

$$R \propto t^q, \quad G \propto t^{-r},$$

where

$$q = 2(1+\omega)/(4+3\omega), \quad r = 2/(4+3\omega) \tag{8.17}$$

and ω is a 'coupling constant' between the scalar field and the geometry. $\omega \to \infty$ gives the Einstein–de Sitter model. Note that for general ω, $G\rho t^2 = $ constant. Dirac's 1937 theory is obtained by setting $\omega = -2/3$.

The Brans–Dicke theory makes slightly different predictions from general relativity for the deflection of light by the Sun and for the perihelion advance of a planet. However, analysis of the lunar ranging data for the 'Nordtvedt' effect (a breakdown in the equivalence principle) gives a limit for $\omega > 29$, which means that the models differ negligibly from those of general relativity (and $-\dot{G}/G < 10^{-12}\ \mathrm{yr}^{-1}$. And if helium is synthesized in the big bang, there is an even stronger limit, $\omega > 100$.

In the Hoyle and Narlikar and Dirac (1973) theories the light deflection and perihelia advance tests would give the same results as general relativity. The variation of G arises because there are two time-scales, atomic time and cosmological time, which no longer coincide.

8.6 Anisotropic and inhomogeneous universes, chaotic cosmology

The fact that isotropy together with the Copernican principle (that we are not in a special place) implies homogeneity, and the high degree of isotropy of the microwave background radiation, made it natural to investigate isotropic and homogeneous models. However, the universe is clearly inhomogeneous on mass-scales up to at least those of galaxies and clusters and we have seen in Chapter 6 that significant density irregularities must have been present at very early times. We can, of course, take the rather supine view that the universe was created with a smooth and isotropic structure on the large scale, but with just sufficient structure on the small scale to evolve to the universe of galaxies and clusters that we see now. A more interesting hypothesis is to suppose that the universe began in a state of regularity apart from small and inevitable 'statistical' fluctuations. We might then hope that these could, in time, grow to form galaxies and clusters. This point of view has been called 'quiescent' cosmology.

A more dramatic speculation was put forward by Misner and has acquired the name of 'chaotic' cosmology: the present universe would have arisen *whatever the initial conditions.* The idea is that no matter how anisotropic and inhomogeneous the universe was initially, gravitation and other physical processes would have caused it to evolve to the present state, after a long enough time.

The mechanism originally suggested by Misner to smooth out the initial non-uniformity in the matter distribution was neutrino viscosity, but it has now been shown that this cannot remove arbitrary anisotropy nor any inhomogeneity over a scale of 1 $M_{\odot}$. A more promising mechanism for damping out the anisotropy has been advocated by Zeldovich and other Soviet cosmologists. The rapidly changing gravitational field during the very earliest moments after the big bang (10^{-43} s $< t < 10^{-23}$ s) leads to spontaneous creation of particle pairs and radiation at the expense of gravitational energy (analogous to Hawking radiation from black holes—see Section 6.6). Anisotropy and inhomogeneity might be smoothed out because of the additional creation occurring in regions where the energy is higher. Unfortunately there appears to be no experimental means of testing such ideas.

An objection that has been raised to chaotic cosmology is that the erasure of an arbitrary amount of primordial chaos must be paid for by an appropriate quantity of radiation entropy. This is measured by the number of photons per baryon, $\sim 10^9$ at the present epoch, which seems a very low value in this context.

8.7 Grand unified theories and inflation

The success of the Weinberg–Salam unification of the weak and electromagnetic interactions based upon the 'SU(2) $\times$ U(1)' group and the development of the quark theory of strong interactions based on the group SU(3) has led Pati and Salam, Georgi and Glashow, and others, to propose 'grand' unified theories to unify the strong, weak, and electromagnetic interactions. Even more ambitious 'theories of everything' which unify all these interactions with gravity are also being investigated.

The grand unified theories, of which the simplest is described by the structure of the SU(5) group, have some interesting consequences for cosmology. These theories attempt to place quarks on the same footing as electrically charged particles like the electron, and for this to be possible quarks must be able to change into electrons and positrons. Because of this possibility the proton can decay by transformation of its three interior quarks into a positron and a pion. This takes about 10^{31} years on average but by studying a large mass of material it is hoped to test this in a few years time. This non-conservation of baryons requires the existence of superheavy particles (either 'gauge' particles or 'Higgs bosons') of mass $\sim 10^{15}\,\mathrm{GeV}/c^2$. In the very early universe ($T > 10^{28}$ K, $t < 10^{-35}$ s) when the energy of a photon is comparable with the rest-mass energy of these superheavy particles, the baryon non-conserving processes can be very important and can convert an initial universe with an arbitrary level of baryon asymmetry into one with just the excess of baryons that we see today, 1 per 10^9 photons. At the epoch when baryons and antibaryons annihilate ($T \sim 10^{12}$ K), we are left with a universe composed predominantly of baryons as observed.

In Section 5.4, I gave a brief account of the *inflationary scenario*, in which the universe goes through a phase of prolonged exponential expansion during the very early universe. Because this solves the horizon and flatness problems, this scenario is very popular with cosmologists and particle physicists, though it is hard to see how it can be tested.

8.8 Eddington's magic numbers

Eddington remarked that if you work out the ratios

$$\frac{\text{electromagnetic force between proton and electron}}{\text{gravitational force between proton and electron}} = \frac{e^2}{Gm_\mathrm{e}m_\mathrm{p}} = 0.23 \times 10^{40},$$

and

$$\frac{\text{radius of universe}}{\text{classical electron radius}} = \frac{c\tau_0}{e^2/m_\mathrm{e}c^2} \sim 8 \times 10^{40}$$

you have two huge dimensionless numbers of the same order of magnitude and that it would be surprising if this were a coincidence. If $G\rho_0\tau_0^2 \sim 1$, then a third dimensionless number can be deduced:

$$\left\{\frac{\rho_0(c\tau_0)^3}{m_\mathrm{p}}\right\}^{1/2} \sim 10^{40},$$

which is roughly the square root of the number of particles in the universe, if $k = 1$.

Two interpretations are possible. (1) We are at a special epoch, at which these numbers happen to be the same. It has been suggested, for example, that we *are* at a special epoch, namely, the one at which life has evolved to the state that these questions can be asked. Certain conditions are necessary to produce life and these involve relationships between the fundamental constants (this has been called the 'anthropic principle' and was first proposed by Dicke). (2) These numbers remain equal at all epochs, so that some of the fundamental constants change with time (Dirac's large number hypothesis). In Dirac's 1937 theory

$$G \propto t^{-1}, \quad R \propto t^{1/3}, \quad \tau \propto t, \quad \text{and } G\rho t^2 = \text{constant},$$

so all three dimensionless numbers are proportional to t and remain in the same proportions at all times. However, this leads to too short an age for the universe ($\tau/3$). Gamow has suggested $e^2 \propto t^{-1}$ ($G = $ constant), as an alternative, but this runs into difficulties with isotope abundances and with the fact that the lines in quasar and galaxy spectra show that the *fine-structure constant* e^2/hc is not varying with epoch.

In Dirac's 1973 theory

$$G \propto t^{-1}, \quad R \propto t, \quad \tau \propto t, \quad \text{and } G\rho\tau^2 = \text{constant},$$

and there is the additional feature of continuous particle creation, either uniformly as in the steady state cosmology, or in proportion to existing matter. This rate of variation of G is beginning to be in conflict with existing observation limits (Section 8.5).

8.9 Problems

8.1 Work out the diameter–redshift relation and the integrated background radiation in (a) the de Sitter model, (b) the steady state model.

8.2 For $0 \le t \le t_1$ there is a radiation-dominated universe ($k = \Lambda = 0$) with $\rho_r = aT^4$, $T = T_1(R_1/R)$. For $t_1 < t < t_2$ there is an inflationary epoch in which $k = \rho = 0$, $\Lambda = 8\pi G\, \rho_r(t_1)$. For $t_2 < t \le t_3$, we again have a radiation-dominated universe ($k = \Lambda = 0$) with $\rho_r = aT^4$, $\rho_r(t_2) = \rho_r(t_1)$ and $T = T_0\ (R_0/R)$. Finally for $t_3 < t \le t_0$ (the present epoch), we have a matter-dominated universe ($k = \Lambda = 0$).
Assuming that $t_1 = 10^{-35}$ s, $t_2 = 10^{-33}$ s, $t_3 = 3 \times 10^5$ yrs, $t_0 = 10^{10}$ yrs (1 yr $= 10^{7.5}$ s), give the behaviour of $R(t)$ for $0 < t < t_0$, including a sketch (not to scale) of log R versus log t. Give a sketch also of log T versus log t. Estimate the redshift factor $Z = R_0/R$ for each of the epochs t_i, $i = 1, 2, 3$.

Epilogue: Twenty controversies in cosmology today

In this epilogue I outline some of the cosmological controversies and issues not yet totally resolved. It is a personal perspective, as of March 1996. In the twenty years since I wrote the first edition of this book, there has been progress in almost all of the areas I listed as controversial then. A few controversies have, in my opinion, been resolved, to be replaced by new issues. For most others the area of debate has shifted. Other cosmologists might consider many of these questions settled, or suggest other items for inclusion. But I believe it is important to emphasize that science is to do with argument and controversy and is not a sacred canon of 'knowledge' handed on from generation to generation.

1 General relativity

Most of the models of the universe described in this book are based on the general theory of relativity. In the past two decades great effort has gone into testing the predictions of general relativity, mainly by tests within the solar system. The theory has survived these tests remarkably successfully, to an accuracy of 1 per cent or better. The Brans–Dicke theory is now forced to be so similar to general relativity that it has lost its interest as a serious rival theory.

Since general relativity has been so successful in solar system applications, where the gravitational field is relatively weak, it is important that the theory should be tested in a stronger field. There are now several good candidates for black holes in binary systems in our Galaxy, most detected as X-ray sources. The first good candidate found was Cygnus X-1, whose X-rays are believed to originate from gas heated by falling on to an accretion disc surrounding a 30 $M_{\odot}$ black hole. The X-rays show rapid flickering on a time-scale of 1 sec. Two other good candidates are the X-ray transient source V616 Mon, and LMC X-3 in the Large Magellanic cloud. Massive black holes, with masses in the range 10^6–10^9 $M_{\odot}$, are postulated to lie in the nuclei of active galaxies and quasars, and to be responsible for the acceleration of relativistic particles in these sources. Good dynamical evidence has been given for compact massive objects in the nuclei of M87 using the Hubble Space Telescope and of NGC4258 using radio observations of water maser lines.

The next generation of gravitational wave detectors may be capable of detecting gravitational radiation from the stellar collapse which takes place during supernova explosions and from the coalescence of closely orbiting neutron stars. Gravitational radiation from close binary systems containing neutron stars should be clearly detected from ESA's proposed space-borne gravitational wave experiment LISA. The existence of gravitational radiation has already been deduced indirectly from the orbital evolution of the 'binary pulsar', a binary system in which one of the stars is a pulsar.

The advance of experimental gravitation in the past decades and the successes of general relativity in these experiments do not reduce the need for continuing effort. In fact it is the theories without effective rivals that require the most vigilant testing.

Perhaps the most likely way in which general relativity might be proved wrong is through variation in the gravitational constant G with time. Forthcoming laboratory experiments may soon be able to improve on the accuracy already achieved in lunar laser ranging experiments, in studies of lunar and planetary motion, and from paleontological arguments. These limits are on the brink of a convincing refutation of the most worthwhile variable G models, those of Dirac (Section 8.4).

2 Quantum gravity and the initial singularity

Near the initial singularity the space–time curvature becomes very large and quantum mechanical effects become significant. The general theory of relativity breaks down and needs to be replaced by a quantum theory of gravity. The time when this happens is the Planck time, $t_{\rm Pl} \sim (Gh/c^5)^{1/2} \sim 10^{-43}$ s, and at these times we have no adequate theory to guide us. There is no agreement yet on the best way to quantize gravity, but the superstring model of Green and Schwartz is attracting much theoretical interest and effort.

Between the Planck time, which corresponds to the radius of the universe being equal to its Compton wavelength h/mc, and the Compton time, when the radius of the universe is equal to the Compton wavelength of a proton ($\sim 10^{-23}$ s), quantum-gravitational effects like particle creation (Section 8.6) are still important but the concepts of the gravitational field and the metric of space–time can be used. Much theoretical effort has gone into the study of this era, particularly in the framework of chaotic cosmology (Section 8.6), one of the motivations for this effort being the hope that a quantized theory of gravity can eventually be developed.

Grand unified and supersymmetric theories (Section 8.7) hold out the exciting prospect of unifying the large and small scale, but this excitement has to be tempered with an awareness of how far we are extrapolating from astronomical observations. Our direct evidence, the microwave background, comes from an epoch 10^5–10^6 years after the big bang. Our indirect evidence, the primordial abundance of helium and deuterium, reaches back to about one second after the big bang. When we talk about the Planck era we are extrapolating the expansion of the universe backwards by a further 40 orders of magnitude in time. However, the discovery of microwave background fluctuations by COBE does hold out the hope of at least constraining ideas about the very earliest instants of the universe.

3 The origin of the motion of the Local Group with respect to the microwave background

The discovery in 1977 by several groups working with airborne and balloon-borne experiments that our Galaxy, and the entire Local Group of galaxies, are moving through the cosmic frame at a speed of about 600 km s^{-1} (see p. 93) was at first surprising. The speed appeared to be high compared with that deduced at the time for the random motions of other groups and clusters of galaxies. And the direction did not agree with that deduced from studies of the anisotropy in the Hubble flow of nearby galaxies.

In 1991 three-dimensional maps of the galaxy distribution made by my collaborators and myself from the IRAS (Infrared Astronomical Satellite) sky survey showed that the motion of the Local Group could be accounted for by the net attraction of the galaxies and clusters within about 100 Mpc. The direction of the net attraction of these galaxies and clusters agreed well with the direction of the Local Group's motion. And the speed of our motion could be accounted for provided the density of the universe is close to the critical value (i.e. the density parameter $\Omega_0 \approx 1$).

Although several different groups have reached a similar conclusion both about the origin of the Local Group's motion and the value of Ω_0, there is still controversy on this issue. Some groups argue, mainly from studies using clusters of galaxies as a probe, that a significant part of the Local Group's velocity is caused by structures at much greater distances than 100 Mpc. For example, a concentration of rich clusters at a distance of about 150 Mpc in the direction of Centaurus discovered by Shapley has been cited as a potential attractor for the Local Group.

There have also been numerous claims of anomalous streaming motions, where galaxies in particular regions of the sky are moving together at unexplained high velocities. As studies of these regions have improved, discrepancies between observed peculiar velocities and the density structures giving rise to them have tended to be resolved.

Deeper galaxy redshift surveys, now in progress, should settle these issues.

4 Isotropy of the universe

The limits on anisotropy of the microwave background radiation, apart from the dipole anistropy discussed above, 0.001 per cent at the moment, place severe limits on the anisotropy of the universe. It remains a matter of controversy *why* the universe should be so isotropic. Within the framework of general relativity, a first answer appears to be that it is simply a consequence of the initial conditions in the universe. Mach's principle, to which much importance has been attached by supporters of the steady state theory, and which is a major part of the motivation for the Brans–Dicke theory, was shown by Gödel not to be automatically incorporated into general relativity. This principle relates the local inertial frame to the large-scale distribution of matter in the universe and would forbid the arbitrary relative rotation between these found by Gödel to be permissible in general relativity.

In Misner's chaotic cosmology the universe started off highly anisotropic and inhomogeneous but somehow evolved to isotropy and homogeneity on the large scale (Section 8.6). Hawking, on the other hand, argues that if the universe were not isotropic we would not be here, since galaxies would not be able to form in an anisotropic universe. Thus the universe is isotropic *because we are here.*

Inflation appears to provide a partial solution to the problem of the isotropy of the universe. The inflation of a small region by tens of orders of magnitude (more than a factor 10^{28}) tends to reduce any anisotropy in this region by a huge factor. Presumably, though, we cannot start from an arbitrary degree of anisotropy and isotropize it by inflation.

The existence of an isotropic frame of reference at every location in the universe is also paradoxical from the viewpoint of general relativity. General relativity tells us that at each location we can adjust our acceleration to find a frame in which special relativity holds, the local freely falling frame. Within that frame all observers moving with uniform velocity with respect to each other are equivalent, so we have an infinite family of freely falling frames. In cosmology, however, we go further and say that there is a unique frame in which the universe looks isotropic. Not just the acceleration but also the velocity of the frame is fixed. At the moment we have to attribute this, whichever explanation of the isotropy of the universe we adopt, to the particular history of the universe.

5 The flatness problem and the value of Ω_0

If we assume for the moment that the cosmological constant, Λ, is zero (see Section 7 below), then the flatness of the universe is measured by how close Ω_0 is to 1. Observationally, the contribution to Ω_0 from visible stars in galaxies is about 0.01, while the density of baryonic matter inferred from cosmological nucleosynthesis of the light elements is $0.05\ (50/H_0)^2$. Dynamical estimates of Ω_0 from galaxy motions range from 0.2 to 1, with the IRAS galaxy surveys favouring values at the high end of this range. Thus we can be reasonably confident that Ω_0 lies between 0.03 and 2 (allowing for observational uncertainties), Although this seems quite a broad range, encompassing models with both positive and negative spatial curvature, this range does pose a *flatness problem.* When the universe was 1 s old, the value of Ω would have been within one part in 10^{15} of unity, and at the inflationary era 10^{-35} s after the big bang, $|\Omega - 1| < 10^{-50}$. How did our universe come to be so extraordinarily flat at these early times? Another way of putting this is as an age problem. Had the universe not been so flat, it could never have survived this long. Either the universe would have recollapsed long ago, or the average density of matter in the universe would be so low that galaxies and stars could never have formed.

As we saw in Chapter 5, inflation provides an answer to this problem, the period of exponential expansion driving the universe to a very high degree of flatness. Most inflation models assume that the universe would still be very close to flatness today, with $| \Omega_0 - 1 | < 10^{-4}$. However inflation models are not so refined that we can rule out the possibility that the inflation was just not quite strong enough to eliminate the pre-existing curvature. In this case Ω_0 could lie anywhere in the range permitted by observations, though inflation would still be needed to solve the flatness problem.

If inflation did not occur, we have to resort to arbitrarily postulating the appropriate initial conditions, or processes at the Planck era. The statement that the universe has to have had this history in order for galaxies, stars, and ourselves to exist is verging on the tautological and hardly merits the title *anthropic principle* which some authors give it.

6 The horizon problem and inflation

More serious than the flatness problem is the *horizon problem*. When we look at the microwave background radiation in two opposite directions on the sky, we are looking back to regions that, according to the simplest model of the universe, have never been in communication. In fact they are separated by over 70 multiples of the radius of the horizon at that epoch. How then did these regions come to be so similar to each other, satisfying homogeneity and isotropy to one part in 100 000?

Again, inflation provides a solution to this problem, since the pre-inflation patch of the universe which grows to be the observable universe today is so small that it would be well within the horizon at very early times.

If inflation did not occur, then we are again forced back either on appropriate initial conditions or on processes at the Planck era.

7 The cosmological constant, Λ

Einstein introduced the cosmological repulsion in order to generate a static model of the universe. When the expansion of the universe was discovered, this additional force was no longer needed, though it remained a possibility within general relativity. Although it fell from favour for many decades, the cosmological term is staging something of a comeback today. The interpretation of Λ is that it corresponds to the energy-density of the vacuum and particle physics arguments suggest that the natural value for this is an extremely large number. This is taken advantage of in the inflationary scenario, where a phase transition is hypothesized to leave the vacuum in an anomalous state of very high energy-density, which then drives a phase of rapid exponential expansion. The evidence from the statistics of the redshifts of gravitationally lensed quasars suggests that $\lambda_0 = \Lambda/3H_0^2 < 0.95$, about 10^{120} times smaller than the inflationary value. This makes it natural to assume that some process, as yet unspecified, caused Λ to decrease to zero.

A non-zero cosmological constant might, however, have certain advantages.

Some observational arguments favour a value for Ω_0 of about 0.2. A model with $\Omega_0 = 0.2$, $\lambda_0 = 0.8$ would then satisfy the inflationary requirement of a flat (or very close to flat) universe. It would also allow a resolution of the problem of a Hubble time greater than the age of the universe (see Section 8 below). Simulations of such a model show that it is capable of generating a spectrum of large-scale density fluctuations consistent with that seen in galaxy redshift surveys and with COBE.

8 The Hubble constant and the age of the universe

The value of the Hubble constant has continued to be a matter of raging controversy for the past 20 years (see my book *The cosmological distance ladder* (Rowan-Robinson 1985) for some of the history of this). A variety of new methods have come into play to measure the Hubble constant, for example the gravitational lens time-delay and Sunyaev–Zeldovich methods (see Chapter 3). Traditional distance methods like Cepheid variable stars have been pushed out to unprecedented distances, reaching galaxies of the Virgo cluster. Despite all this work the value of H_0 remains highly controversial, though most methods give values in the range 50–80 km s^{-1} Mpc^{-1}.

The age of the oldest stars in our Galaxy is about 13 billion years, with an uncertainty of at least 1 billion years either way. Radioactive dating of long-lived isotopes in the solar system also gives ages for the Galaxy of 10–15 billion years.

If $H_0 = 50$, then an Einstein–de Sitter universe ($k = 0$, $\Lambda = 0$) would yield an age of the universe of 13 billion years, consistent with the age of our Galaxy. On the other hand if $H_0 = 80$, such a model would give an age of only 8 billion years, clearly too short. In this case we would probably be forced to the model with $\Omega_0 = 0.2$, $\lambda_0 = 0.8$, for which the age would be 14 billion years.

Clearly it is vital to establish the Hubble constant to a better accuracy and to reexamine the methods for determining the age of the Galaxy.

9 Large-scale homogeneity and large-scale structure

The combination of the microwave background studies by the COBE satellite and by ground-based and balloon-borne experiments, and of large-scale galaxy redshift surveys both at optical and far-infrared wavelengths, has begun to give us a good understanding of the amplitude of density fluctuations on different scales. On the largest scales amenable to studies of the galaxy distribution, about 100 h^{-1} Mpc, the characteristic rms density variation is a few per cent, so we can now say that the galaxy distribution is homogeneous on large scales to at least this level. On the even larger scales studied by COBE, an even higher degree of homogeneity can by inferred.

The spectrum of density fluctuations on different scales should allow us to deduce something about the nature of the dark matter in the universe and the cosmological parameters. However, at the moment there appear to be a wide range of possibilities, including: mixed hot and cold dark matter, cold dark matter with a tilted initial density perturbation spectrum, hot dark matter with cosmic strings, and low-Ω cold dark matter with non-zero Λ.

An important issue is whether the initial spectrum of density fluctuations is Gaussian, as expected in inflationary scenarios, or non-Gaussian, as might be expected if topological defects like cosmic strings play a major role. Are the voids, sheets, and filaments in the galaxy distribution a natural consequence of models in which structure grows purely under the action of gravity? Are the claimed periodicities in the redshift distribution in deep pencil-beam surveys towards the galactic poles real?

10 Antimatter

In our vicinity there is a strong preponderance of matter over antimatter, and it is normally assumed that this is true for the universe as a whole, as a result of initial conditions. Why should this be so? Some cosmologists, especially Alfvén and Omnes, have argued that there are equal amounts of matter and antimatter in the universe, but that they remain spatially segregated. The encounter of lumps of matter and antimatter (whole galaxies or clusters perhaps) would yield enormous amounts of annihilation energy. When annihilation occurs, pions (π^0) are produced, which decay to give gamma rays. A peak in the gamma-ray background would be expected at 100 MeV, but this is not observed. Cosmic-ray and gamma-ray data exclude the possibility of large amounts of antimatter from the solar system, the Galaxy, the Local Group, galaxy clusters, and the intergalactic medium (if any exists).

Grand unified theories (Section 8.7) offer the possibility of explaining the excess of matter over antimatter which appears to prevail in the universe at the present epoch.

11 Primordial helium and deuterium

According to the standard hot big-bang picture about 24 per cent by mass of the material from which galaxies formed was in the form of helium (^{4}He) and about 0.003 per cent was in the form of deuterium (^{2}H), both produced in thermonuclear reactions in the fireball (Section 5.3). Since these abundances provide an important confirmation for the big-bang picture it is important to be sure that these elements could not have been produced in other astrophysical processes.

The production of a 24 per cent abundance of helium in stellar processes would be accompanied by an energy release so great that it could only be identified with the microwave background radiation. Some old stars in our Galaxy with very weak or non-existent helium lines, which could have pointed to a non-primordial origin for helium, are now understood in terms of normal astrophysical processes which have segregated their helium out of sight.

Only very unusual scenarios can give significant deuterium production, for example huge fluxes of very high-energy cosmic rays at epochs soon after recombination. Normally any deuterium which is made is immediately dissociated or converted to helium. Primordial deuterium would be destroyed in the hot interiors of stars so the present observed abundance is likely to be lower than the primordial abundance.

The abundances of helium and deuterium produced during the fireball can be modified by varying G, anisotropic expansion, additional unknown elementary particles, or neutrino degeneracy, and this has been used to place strong limits on all these effects.

12 Formation of galaxies

There has been considerable progress in modelling how galaxies form from an initial spectrum of density fluctuations using n-body computer simulations. Several scenarios (cold dark matter, mixed dark matter, low-Ω cold dark matter with non-zero Λ) are capable of generating simulated galaxy distributions which are quite similar to the observed galaxy distribution on the sky. Simulations which incorporate hydrodynamics are beginning to be able to model the formation of individual galaxies. However, these simulations do not give much guidance on the epoch at which star formation and heavy element production begins in galaxies, or how the different types of galaxy are formed.

Direct searches for protogalaxies have had a rather mixed success. Many examples of quasars and some examples of radio galaxies have been found with redshifts in the range 3–4.9. The optical and near-infrared continuum of some of the radio galaxies can be modelled in terms of a relatively recent (10^8–10^9 yrs) burst of star formation. A few examples of very high luminosity ($> 10^{14}$ solar luminosities) far-infrared galaxies at $z = 1$–2.5 have been found, with some of the more extreme objects being clearly amplified by gravitational lensing. These are good candidates for being in the process of undergoing a major episode of star formation. High-redshift galaxies have also been found associated with high-redshift quasars and with absorption line systems in quasars. Direct optical and near-infrared surveys for protogalaxies, on the other hand, have been unsuccessful, perhaps because star-forming regions quickly become shrouded in dust.

Although it seems clear that the main star-formation in galaxies takes place between epochs corresponding to $z = 1$–5 and that interactions and mergers play a major role, the details of how galaxies form remain very unclear.

13 Is there an intergalactic medium?

The very accurate Planckian form of the microwave background spectrum found by COBE eliminates the possibility of a hot intergalactic gas at the critical density, which had been postulated as a possible source of much of the X-ray background. A lower density of hot gas could make some contribution to the X-ray background.

The damped Lyman-α clouds seen in quasar absorption line systems probably represent the progenitors of galactic discs. At $z \approx 3$, they contribute about $\Omega = 0.003$ to the density of the universe. The Lyman-α forest clouds, which have much lower column density, could be primordial clouds, but contribute an even lower figure to Ω.

14 Dark matter

Since the contribution of visible stars to the average density of the universe is only $\Omega_0 \approx 0.01$, whereas the baryonic matter contributes $\Omega_b \approx 0.05(50/H_0)^2$, there must be baryonic dark matter in galaxies. This could be in the form of low-mass stars, white dwarfs, brown dwarfs, Jupiters or black holes. There is also direct dynamical evidence for dark halos surrounding all types of galaxy, either through their rotation curves or velocity dispersions. Some low-mass stars have been detected in the halo of our Galaxy through microlensing effects, but they are unlikely to contribute more than about 30 per cent of the mass of the halo.

Galaxy formation scenarious require the existence of primordial density perturbations in non-baryonic dark matter. It is therefore likely that the halo of our Galaxy is primarily non-baryonic, probably some form of cold dark matter. The most popular candidates are the neutralino, the lightest supersymmetric particle, or the axion. Underground searches for the neutralino are in progress and should be in reach of the predicted event rates within a few years.

Hot dark matter, predicted to make up 20 per cent of the mass of the universe in the mixed dark matter model and 95 per cent in the hot dark matter with cosmic strings scenario, would probably take the form of a neutrino with a mass of 3–30 eV/c^2, probably the tau neutrino. The CHORUS and NOMAD experiments at CERN should be capable of measuring such a mass for this neutrino within a few years.

15 Optical counts of galaxies

Optical and near-infrared counts of galaxies show that at faint magnitudes ($B = 23$–28 m.) there is an excess population of blue galaxies, probably dwarf galaxies at $z \approx 0.5$–1. These could form a population which at the present epoch is hard to detect in our locality, for example because they are of low surface brightness. Alternatively they may have interacted with larger galaxies and merged with them, so that they no longer have a separate identity today.

A resolution of the nature of the faint blue galaxy population is likely to be important for our understanding of the galaxy formation process.

16 Evolution of quasars, AGN and starburst galaxies

The steep radio source counts show that there is strong evolution in the radio galaxy and radio-loud quasar populations. Redshift surveys of optically selected quasars give direct evidence that the quasar population is strongly evolving. In both cases the evolution is approximately of the form that the characteristic luminosity of the populations was about 30 times greater at redshift 2–3 than it is today. Similar evolution is also seen in X-ray quasars and, perhaps more surprisingly, in the starburst galaxy population (at far-infrared and radio wavelengths).

The explanation of this evolution is not fully known, but it is possible that interactions and mergers between galaxies play a strong role, both in generating bursts of star formation and in channelling gas to a black hole in the nuclei of active galactic nuclei (AGN).

17 Nature of X-ray and γ-ray background

About 30–50 per cent of the X-ray background and perhaps all of the γ-ray background is likely to be generated by quasars and AGN. X-ray sources in rich clusters of galaxies probably contribute no more than 10 per cent of the X-ray background. The bulk of the remainder of the X-ray background appears to be due to a population of emission-line X-ray galaxies, which resemble starburst galaxies in most respects. At soft X-rays there is a significant foreground from local hot (10^6 K) gas, probably generated in supernova remnants.

18 γ-ray bursts

γ-ray bursts were discovered in the 1970s by the US Vela satellites, which had been designed to monitor atmospheric nuclear explosions. The bursts last from a fraction of a second up to a few hundred seconds and over 2000 have now been detected. The bursts never repeat and in no case has the origin of the burst been identified. The distribution on the sky is approximately isotropic, with no concentration towards the Galactic plane or the centre of our Galaxy. The source-count slope, $d \log N / d \log S$, is significantly flatter than the Euclidean slope of -1.5.

There are many theories of their origin, ranging from a solar system population to a distant cosmological one. The isotropy favours a cosmological distribution, in which case the events require an energy of 10^{44} W, which could possibly be generated by the collision of two neutron stars.

19 Solar neutrino problem

Four separate neutrino experiments have all detected electron neutrinos from the Sun at a rate significantly lower than predicted by the standard solar model, typically by a factor of about 2. A possible explanation is that the electron neutrino undergoes oscillations, converting to a muon neutrino, and that the electron and muon neutrinos therefore have non-zero masses, of order a millionth and a few thousandths of an eV/c^2, respectively. This could fit in well with scenarios in which the tau neutrino has a mass of 3–30 eV/c^2 and is a major constituent of the dark matter in the universe.

It is clearly vital that the solar neutrino problem be resolved, since it casts doubts on the credibility of our stellar structure models. Future neutrino experiments which will also be sensitive to muon neutrinos could help resolve the problem.

20 Life in the universe

The antigeocentric viewpoint, which encourages us to believe the cosmological principle, leads naturally to the idea that we are not unique in the universe. Calculations of the probability of other inhabited planets in our Galaxy are rather meaningless at this stage of our knowledge of the origin of life. (An estimate that is often quoted is that the number of technical communities in our Galaxy now is equal to 10 per cent of the average survival time of a technical civilization in years.) But in the framework of the cosmological principle we should assume that there is at least one inhabited planet per galaxy. Naturally it is intriguing to think that somewhere else all the controversies outlined in this epilogue have been resolved. Yet we should not expect to learn these solutions through some intergalactic broadcasting service, for in the framework of evolutionary cosmology, in which galaxies form simultaneously and evolve in parallel, the light from even the nearest comparable galaxy, M31, set off 2 million years ago, long before Andromeda-people would have appeared on Andromeda-Earth. We have to solve these problems on our own.

Answers to problems

Chapter 1

1.1: 9.6×10^{-7} W m^{-2} sr^{-1}, 4.02×10^{-14} W m^{-3}, 2.5×10^{34} W m^{-3}, 2.6×10^{12} K.
1.2: 4.84×10^{-11} W m^{-2} sr^{-1}, 0.00005.

Chapter 2

2.1: 6.3×10^{14}, 2.5×10^{12}, 4.0×10^{7}, 1.6×10^{5}.

Chapter 3

3.1: (a) 0.37, (b) 0.75, (c) 2.16. 3.2: (i) 0–0.03, 0.37–0.85, 2.23–3.11, (ii) 0.03–0.37, 3.11–4.81, 4.81–6.40.

Chapter 4

4.1: 1.3×10^{10} years; $50 \leq H_0 \leq 68$ km s^{-1} Mpc^{-1}

Chapter 5

5.2: (a) $t_2 = 75$ s, $t_1 = 3$ s, (b) $t_2 = 45\,000$ s, $t_1 = 4000$ s. $X(\mathrm{He}) = 0.30$ (a), 0 (b). Case (a) is close to observed primordial helium abundance.

Chapter 7

7.1: see Table 7.2

Chapter 8

8.1: see Table 7.2; Q8.2: $Z_3 = 10^3$, $Z_2 = 10^{26}$, $Z_1 = 10^{56}$.

Further reading

Introductory

Harrison, E. R. (1981). *Cosmology, the science of the universe.* Cambridge University Press.

Hawking, S. (1988). *A brief history of time.* Bantam Press, London.

Hubble, E. (1958). *Realm of the nebulae.* Dover, New York.

Rowan-Robinson, M. (1979). *The cosmic landscape.* Oxford University Press.

Rowan-Robinson, M. (1993). *Ripples in the cosmos.* Freeman, New York.

Silk, J. (1979). *The big bang.* Freeman, New York.

Smoot, G. (1993). *Wrinkles in time.* Little and Brown, London.

Thorne, K. S. (1994). *Black holes and time warps: Einstein's outrageous legacy.* Picador, London.

Weinberg, S. (1977). *The first three minutes.* Basic Books, New York.

At this level

Berry, M. (1977). *Principles of cosmology and gravitation.* Cambridge University Press.

Sciama, D. W. (1971). *Modern cosmology.* Cambridge University Press.

More advanced

Coles, P. and Lucchin, F. (1995). *The origin and evolution of cosmic structure.* Wiley, New York.

Kolb, E. W. and Turner, M. S. (1990). *The early universe.* Addison-Wesley, Reading, Mass.

Narlikar J. V. (1983). *Introduction to cosmology.* Jones and Bartlett, Boston.

Peebles, P. J. E. (1993). *Principles of physical cosmology.* Princeton University Press, Princeton, NJ.

Rowan-Robinson, M. (1985). *The cosmological distance ladder.* Freeman, New York.

Weinberg, S. (1972). *Gravitation and cosmology.* Wiley, New York.

See also many excellent review articles in *Annual Reviews of Astronomy and Astrophysics*, Palo Alto, California.

Glossary

baryon: type of hadron (q.v.), consisting of proton, neutron, and the unstable hyperons (and their antiparticles).

big-bang models: expanding universe models in which the density was infinite at a finite time in the past.

blackbody radiation: a perfectly efficient radiator or absorber of radiation is called a black body. When matter and radiation are in complete thermal equilibrium, e.g. during the fireball, the radiation will have a blackbody, or Planck, spectrum. The background radiation at microwave frequencies has the spectrum of a 2.7 K black body and is believed to be a relic of the fireball.

black hole: a region in which the matter has collapsed together to such an extent that light can no longer escape from it.

bremsstrahlung: radiation from an ionized gas due to electrons moving in the electrostatic field of ions.

Cepheids: stars whose brightness varies sinusoidally on a period of 2–100 days, the period being directly related to the mean luminosity of the star.

co-moving observer: one who is at rest with respect to the substratum.

Compton scattering: the scattering by free electrons of photons which, in the rest-frame of the electron, have energies greater than the rest-mass energy of the electron ($m_e c^2$). When the scattering results in a boost of the photon energy, it is often called inverse Compton scattering.

Copernican principle: the Earth does not occupy a privileged position in the universe.

cosmic rays: relativistic particles, both nuclei and electrons, continuously bombarding the Earth. Some come from the Sun, others from pulsars, supernovae, and other violent events.

cosmological principle: the universe as seen by fundamental observers is homogenous and isotropic.

cosmological time: in a homogeneous universe the proper times of fundamental observers can be synchronized to give a universal, cosmological time.

critical density: that average density of the universe which divides big-bang models that will keep on expanding for ever from those that will ultimately recontract.

deceleration parameter: measures the rate at which the expansion of the universe is slowing down, in dimensionless form. Must be positive if gravity is the only force acting.

diameter distance: distance deduced assuming the apparent angular size of an object falls off inversely with distance.

distance modulus: the difference between the apparent and absolute magnitudes of an object, equal to 5 lg (distance in pc) – 5.

entropy: the degree of disorder in a system.

epoch of decoupling: the moment in the evolution of a big-bang universe when the matter recombines and becomes transparent to radiation.

field equations: differential equations which in the general theory of relativity relate the geometry of space–time, described by the metric, to the distribution of matter and other forms of energy.

fireball: the phase in a big-bang universe prior to the epoch of decoupling, when the matter is completely opaque to radiation, and matter and radiation are in thermal equilibrium.

flux: the total energy received from a source per sec per unit area normal to the direction of the source (unit: W m^{-2}).

flux density: the flux per unit bandwidth (unit: W m^{-2} Hz^{-1}). Also known as the monochromatic flux.

fundamental observer: observer who is at rest with respect to the substratum, i.e. co-moving with it.

Galactic coordinates (l, b): spherical polar coordinates analogous to latitude and longitude, in which the plane of the Galaxy plays the part of the equator, and the direction of the Galactic centre corresponds to zero longitude.

gravitational wave: according to general relativity waves can be transmitted via the gravitational field, just as light is transmitted via the electromagnetic field.

hadron: the heavier elementary particles (protons, neutrons, mesons) which take part in strong nuclear interactions.

heavy elements: all elements apart from hydrogen, helium, and the light elements lithium, beryllium, and boron, are referred to as heavy elements by astronomers. On average they make up about 1 per cent of the matter in our Galaxy.

helium flash: the moment in a star's evolution when the core, exhausted of hydrogen, has heated up sufficiently for helium fusion to begin.

Hertzsprung–Russell (HR) *diagram:* plot of stellar luminosity against surface temperature (or, in practice, colour or spectral type), in which the evolution of stars of different masses may be followed.

homogeneity: a homogeneous universe is one that appears the same to all fundamental observers.

horizon: surface which divides those particles we can already have observed from those we cannot yet know anything about.

Hubble distance: the distance at which galaxies would have a redshift of 1 on the basis of the Hubble law. Current value is 6000 Mpc or 2×10^{10} light years.

Hubble law: the redshift of objects in the universe increases linearly with distance. Most cosmological models predict departures from linearity when the redshift is no longer $\ll 1$.

Hubble parameter (or constant): the slope of the redshift–distance relation (unit: km s^{-1} Mpc^{-1}). Current value is 50–80 km s^{-1} Mpc^{-1}.

Hubble time: the time for the universe to double its size expanding at the present rate, if the redshift is interpreted as due to recession. Current value is 2×10^{10} years.

hydrogen burning: nuclear fusion of hydrogen into helium in the interior of stars.

H_{I} cloud: cloud of cool, neutral hydrogen.

H_{II} region: cloud of hot, ionized hydrogen, usually heated by a luminous star.

inertial frame: frame of reference in which Newton's first law of motion holds.

intensity: the (monochromatic) intensity of an extended source of radiation is the flux (density) per unit solid angle (unit: W m^{-2} sr^{-1} Hz^{-1}).

interval: measured in an inertial frame, the interval between two events separated by distance dr and time dt is defined by $ds^2 = dt^2 - dr^2/c^2$, so that it is zero if the events can be connected by a light signal.

invariant: scalar quantity that has the same value in all frames of reference.

isotropy: an isotropic universe is one which to a fundamental observer looks the same in every direction on the sky.

Jansky (Jy): astronomical unit of flux-density: 1 Jy = 10^{-26} W m^{-2} Hz^{-1}.

leptons: the lighter elementary particles (neutrinos, electrons, muons) which do not take part in strong nuclear interactions.

light year: distance travelled by light *in vacuo* in one year: 1 light year = 9.46×10^{15} m.

Local Group: small group of 30 or so galaxies of which our Galaxy is one of the dominant members (see Table 1.1; p. 4).

luminosity: the bolometric or total luminosity is the energy emitted by a source per unit solid angle per second (unit: W sr^{-1}). The monochromatic luminosity is the luminosity per unit bandwidth (unit: W sr^{-1} Hz^{-1}).

luminosity class: classification of spiral galaxies according to the appearance of their spiral arms, each class having a different mean luminosity. Allows spirals, especially of type Sc, to be used as a distance indicator.

luminosity distance: distance of a source assuming the inverse square law for radiation holds.

magnitude: logarithmic scale of brightness used by astronomers, one magnitude corresponding to a change of 0.4 in lg. The apparent magnitude is -0.4 lg (flux) + constant. Photographic (m_{pg}), visual (m_v), photoelectric (U, B, V) magnitudes refer to magnitudes determined with different detectors and filters. Bolometric magnitude is the magnitude that would be obtained if all the light from the source could be detected. Absolute magnitude, the unit of luminosity, is the magnitude a source would have at a distance of 10 pc.

main sequence: locus in the HR diagram corresponding to the hydrogen-burning phase in the life of stars of different masses.

metric: the relationship between the coordinate difference between two nearby events and the interval; defines the geometry of space-time.

nucleon: nuclear particles, i.e. protons and neutrons.

neutron star: cold, degenerate, compact star in which nuclear fuels have been exhausted and pressure support against gravity is provided by the degeneracy pressure of neutrons.

nucleosynthesis: the building up from hydrogen of the elements in the periodic table by means of nuclear reactions: in the fireball for helium; in stars for the other elements.

parallax: the change in apparent direction of a star due to the Earth's motion round the sun.

parsec (pc): distance at which the radius of the Earth's orbit subtends one second of arc: 1 pc = 3.26 light years.

peculiar velocity: that part of a galaxy's velocity due to its random motion relative to the substratum.

Planck spectrum: the energy distribution characteristic of a black body.

principle of equivalence: gravity may be transformed away locally by choosing a freely falling frame of reference.

proper distance: distance measured by an observer using radar methods.

proper time: the proper time of an observer is the time measured on a clock at rest with respect to him.

pulsar: pulsating radio source associated with neutron star.

quasars: outstandingly luminous quasi-stellar radio sources related to violent events in the nuclei of galaxies.

red giant: phase in star's evolution after completion of hydrogen burning when outer layers become very extended.

redshift: shift in frequency of spectral lines towards the red end of the spectrum, due to recession of source (Doppler shift) or effects of gravity (gravitational redshift). The cosmological redshift (see Hubble law) is usually interpreted as implying expansion of the universe.

relativistic: with velocity close to the speed of light.

scale-factor: the function of time by which all distances scale in a universe satisfying the cosmological principle.

Seyfert galaxy: type of galaxy showing signs of activity in nucleus.

singularity: point where theory predicts that physical variables, especially the density, become infinite. Most notable is the initial moment of big-bang universes, but also happens to matter inside a black hole.

space–time: in relativity theory the three dimensions of space and one of time are treated as a single four-dimensional space–time continuum.

spectral index: the parameter α if the monochromatic luminosity $P(\nu)$ has the form $P(\nu) \propto \nu^{-\alpha}$, where ν is the frequency.

steady state cosmology: model of expanding universe in which all properties are independent of time: continuous creation of matter is necessary to maintain the universe at a constant density.

substratum: the matter in the universe is imagined to be smeared out into a smooth fluid.

synchrotron radiation: radiation from relativistic electrons spiralling in a magnetic field.

Thomson scattering: scattering by free electrons of photons which, in the rest-frame of the electron, have energies much less than the rest-mass energy of the electron ($m_e c^2$). No change in the energy of the electron or of the photon results.

white dwarf: cool, degenerate, compact star, in which nuclear fuels are exhausted and pressure support against gravity is provided by the degeneracy pressure of electrons.

Physical constants and conversion factors

Avogadro constant	L or N_A	$6.022 \times 10^{23}\ \text{mol}^{-1}$
Bohr magneton	μ_B	$9.274 \times 10^{-24}\ \text{J T}^{-1}$
Bohr radius	a_0	$5.292 \times 10^{-11}\ \text{m}$
Boltzmann constant	k	$1.381 \times 10^{-23}\ \text{J K}^{-1}$
charge of an electron	e	$-1.602 \times 10^{-19}\ \text{C}$
Compton wavelength of electron	$\lambda_C = h/m_e c$	$= 2.426 \times 10^{-12}\ \text{m}$
Faraday constant	F	$9.649 \times 10^{4}\ \text{C mol}^{-1}$
fine structure constant	$\alpha = \mu_0 e^2 c/2h$	$= 7.297 \times 10^{-3} (\alpha^{-1} = 137.0)$
gas constant	R	$8.314\ \text{J K}^{-1}\ \text{mol}^{-1}$
gravitational constant	G	$6.673 \times 10^{-11}\ \text{N m}^2\ \text{kg}^{-2}$
nuclear magneton	μ_N	$5.051 \times 10^{-27}\ \text{J T}^{-1}$
permeability of a vacuum	μ_0	$4\pi \times 10^{-7}\ \text{H m}^{-1}$ exactly
permittivity of a vacuum	ϵ_0	$8.854 \times 10^{-12}\ \text{F m}^{-1} (1/4\pi\epsilon_0 = 8.988 \times 10^{9}\ \text{m F}^{-1})$
Planck constant	h	$6.626 \times 10^{-34}\ \text{J s}$
(Planck constant)/2π	$\hbar$	$1.055 \times 10^{-34}\ \text{J s} = 6.582 \times 10^{-16}\ eVs$
rest mass of electron	m_e	$9.110 \times 10^{-31}\ \text{kg} = 0.511\ \text{MeV}/c^2$
rest mass of proton	m_p	$1.673 \times 10^{-27}\ \text{kg} = 938.3\ \text{MeV}/c^2$
Rydberg constant	R_∞	$\mu_0^2 m_e e^4 c^3/8h^3 = 1.097 \times 10^{7}\ \text{m}^{-1}$
speed of light in a vacuum	c	$2.998 \times 10^{8}\ \text{m s}^{-1}$
Stefan–Boltzmann constant	$\sigma = 2\pi^5 k^4/15h^3c^2$	$= 5.670 \times 10^{-8}\ \text{W m}^{-2}\text{K}^{-4}$
unified atomic mass unit (^{12}C)	u	$1.661 \times 10^{-27}\ \text{kg} = 931.5\ \text{MeV}/c^2$
wavelength of a 1 eV photon		$1.243 \times 10^{-6}\text{m}$

1 Å $= 10^{-10}$ m; 1 dyne $= 10^{-5}$ N; 1 gauss (G) $= 10^{-4}$ tesla (T);
0°C $= 273.15$ K; 1 curie (Ci) $= 3.7 \times 10^{10}\ \text{s}^{-1}$;
1 J $= 10^7$ erg $= 6.241 \times 10^{18}$ eV; 1 eV $= 1.602 \times 10^{-19}$ J; 1 $\text{cal}_{\text{th}} = 4.184$ J;
$\ln 10 = 2.303$; $\ln x = 2.303 \log x$; $e = 2.718$; $\log e = 0.4343$; $\pi = 3.142$

Name index

Subject index